S260 Geology
Science: Level 2

The

Block 3
Internal processes

Prepared for the Course Team by David Rothery, Nigel Harris and Andrew Bell

The S260 Core Course Team

David Rothery *(Course Team Chairman and Author)*

Glynda Easterbrook *(Course Manager and Author)*

Iain Gilmour *(Multimedia Development Coordinator and Author)*

Angela Coe *(Block 4 Chair and Author)*

Other members of the Course Team

Gerry Bearman *(Editor)*

Roger Beck *(Reader)*

Andrew Bell *(Author)*

Steve Best *(Graphic Artist)*

Evelyn Brown *(Author)*

Sarah Crompton *(Designer)*

Janet Dryden *(Secretary)*

Neil Edwards *(Multimedia)*

Nigel Harris *(Author)*

David Jackson *(BBC)*

Jenny Nockles *(Design Assistant)*

Pam Owen *(Graphic Artist)*

David Palmer *(Author & Multimedia)*

Rita Quill *(Course Secretary)*

Jon Rosewell *(Multimedia)*

Dick Sharp *(Editor)*

Peter Skelton *(Author)*

Denise Swann *(Secretary)*

Tag Taylor *(Design Co-ordinator)*

Andy Tindle *(Multimedia)*

Fiona Thomson *(Multimedia)*

Mike Widdowson *(Author)*

Chris Wilson *(Author)*

This publication forms part of an Open University course S260 *Geology*. The complete list of texts which make up this course can be found at the back. Details of this and other Open University courses can be obtained from the Student Registration and Enquiry Service, The Open University, PO Box 197, Milton Keynes MK7 6BJ, United Kingdom: tel. +44 (0)845 300 60 90, email general-enquiries@open.ac.uk

Alternatively, you may visit the Open University website at http://www.open.ac.uk where you can learn more about the wide range of courses and packs offered at all levels by The Open University.

To purchase a selection of Open University course materials visit http://www.ouw.co.uk, or contact Open University Worldwide, Michael Young Building, Walton Hall, Milton Keynes MK7 6AA, United Kingdom for a brochure. tel. +44 (0)1908 858793; fax +44 (0)1908 858787; email ouw-customer-services@open.ac.uk

The Open University, Walton Hall, Milton Keynes, MK7 6AA

First published 1999. Reprinted with corrections 2002. Second edition 2007

Edited, designed and typeset by The Open University.

Printed in Europe by the Alden Group, Oxfordshire.

ISBN 978 0 7492 1882 9

2.1

BLOCK 3 INTERNAL PROCESSES

CONTENTS

1 INTRODUCTION TO BLOCK 3

By this stage in the Course, you should be able to make sense of basic geological materials. We hope you can now look at a geological map or cross-section and grasp what this means in terms of the three-dimensional distribution of the rock units, and examine a hand specimen or thin section and make some deductions about its composition and origin. Don't worry if you do not feel confident with that sort of thing yet; there are more opportunities to practise in this Block. However, whereas Blocks 1 and 2 dealt with the 'building blocks' of geology (e.g. maps, cross-sections, minerals and rocks), Blocks 3 and 4 concentrate on geological *processes*, by which we mean the events that happen as a particular situation develops, and on the *environments* in which these occur. You will be using your skills of observation and reasoning, plus examples of geological processes that we can see happening today, to help interpret events in the past and thus deduce what it would have been like had you been there at the time. This, after all, is one of the main aims of geology as a discipline.

This Block deals with the Earth's internal processes. It encompasses processes that occur below the Earth's surface or that are driven from below the surface, whereas Block 4 is concerned with processes and interactions that lie more completely in the surface environment.

Because this is a geology course, most of the internal processes with which we will be concerned happen at relatively shallow depths, rather than being deeper processes that can be studied only by geochemical or geophysical techniques. Specifically, we will concentrate on processes either within the Earth's *crust* or in the topmost part of the underlying *mantle*. If you have studied S102 *A Science Foundation Course* or S103 *Discovering Science*, you should already know those two terms. However, we think it helpful to begin this Block with a brief review of the Earth's structure and the essentials of plate tectonics (something else you will have met in S102 or S103), to provide a context for the processes you will be studying.

2 THE EARTH

2.1 SETTING THE SCENE – SIZE, MASS AND DENSITY OF THE EARTH

The Earth is the largest of four planets in the inner part of the Solar System. These, together with the Earth's Moon, are similar in terms of size, mass and density (and composition) and constitute the **terrestrial planets** (Table 2.1). They are much less massive, though denser, than the giant planets Jupiter, Saturn, Uranus and Neptune. All these planets are about 4.6 billion (4600 million) years old.

Name	Equatorial radius/10^3 km	Mass/10^{24} kg	Density/10^3 kg m^{-3}
Mercury	2.439	0.330	5.43
Venus	6.052	4.87	5.25
Earth	6.378	5.98	5.52
Moon	1.738	0.0735	3.34
Mars	3.394	0.641	3.90

Table 2.1 Basic data for the terrestrial planets. (The Moon is not a planet in the astronomical sense, because it orbits the Earth rather than the Sun; however, it has the geological attributes of a terrestrial planet. For the same reason, one of Jupiter's satellites, Io, is sometimes classified as a terrestrial planet too.)

Question 2.1 The densities of rock types collected near the Earth's surface range from about 2.35×10^3 kg m^{-3} (sandstone) to 3.25×10^3 kg m^{-3} (peridotite). The average density of the Earth quoted in Table 2.1 falls well outside this range. Consider the implications of this and write down *two* different explanations that could account for the discrepancy.

Actually, the Earth's average density is believed to result from a combination of both the effects described in our answer. As depth (and hence pressure) increases, rock becomes compressed. The effect is slight except where particular pressures are reached, causing phase transitions that allow silicate minerals within the mantle to change to denser structures, such as the olivine/spinel phase transition that you met in Section 4.3.2 of Block 2. However, most of the difference between the density of near-surface rocks and the Earth's average density is because the Earth's central part (its *core*) is not made of silicates at all, but is made of iron-rich material $>1 \times 10^4$ kg m^{-3} in density.

All the terrestrial planets are believed to have iron-rich cores, with the possible exception of the Moon. We know far more about the Earth's interior than that of any other planetary body because it has been mapped by studying how **seismic waves** (vibrations triggered by earthquakes or explosions) travel through it. There is not space in this Course to examine the seismic and other geophysical evidence bearing on the Earth's interior. Instead, we limit ourselves to a descriptive review in the next Section*.

2.2 THE EARTH'S INTERNAL STRUCTURE

Viewed on a global scale, the Earth is composed of concentric layers of differing nature, expressed in two contrasting ways. One is by *composition*, the other is by *mechanical properties*.

2.2.1 COMPOSITIONAL LAYERING

The most fundamental compositional distinction within the Earth is between the **core**, which occupies the centre and is iron-rich, and the overlying rocky material. Study of the propagation of seismic waves shows that the core consists of inner and outer parts. The inner core is solid and has a density consistent with pure iron, though comparison with meteorites suggests that it may have roughly 20% nickel mixed in. The outer core is liquid and appears to consist of molten iron diluted by around 5–15% of less dense elements, most likely one or more out of oxygen, sulfur or potassium. Motion of this fluid outer core (which is electrically conducting) is responsible for the Earth's magnetic field.

Above the core lies the **mantle**, which is virtually all solid and makes up most of the remainder of the Earth. The distinction between core and mantle is very clear cut, because the mantle consists of silicate minerals and has a bulk chemical composition equivalent to that of the ultramafic rock type known as peridotite that you met in Block 2 Section 6.3. However, this is not to say that the whole mantle consists mostly of olivine and pyroxene.

❑ We suggested why higher on this page. What is the reason?

■ It is because of phase transitions in which the minerals are compressed to denser structures, which therefore have different mineral names.

Notable phase transitions in the mantle are believed to occur at about 400 km where olivine is compressed to a denser structure described as spinel (Block 2 Section 4.3.2) and at about 670 km where the spinel structure transforms to an even higher density structure called perovskite. However, these phase transitions do not represent changes in chemical composition.

* A fuller account can be found in Block 1 of S267 *How the Earth Works: The Earth's Interior.*

The outermost solid part of the Earth is the **crust**. The distinction between crust and mantle is less fundamental than between mantle and core. Crust and mantle are both predominantly silicate in composition, and differ in that the crust has a higher percentage of silica than the mantle, and so is made of less-dense rock.

❑ What is the percentage of SiO_2 in an ultramafic rock type such as peridotite?

■ Ultramafic rocks are defined as consisting of <45% SiO_2 (Block 2 Table 6.4).

In fact, mantle composition is thought to be approximately 44–45% SiO_2, whereas the average SiO_2 content of the crust is about 10% higher. The picture is complicated because the Earth has two distinct types of crust, of differing compositions. **Oceanic crust** (*c*. 49% SiO_2) forms the floor of the deep oceans. **Continental crust** (*c*. 60% SiO_2) makes up, as you might expect, the continental landmasses and also the floors of the shallow **shelf seas** that cover the **continental shelves** around edges of most continents. The Earth's layered structure is summarized in Figure 2.1, and the compositions of the mantle and crust are given in Table 2.2.

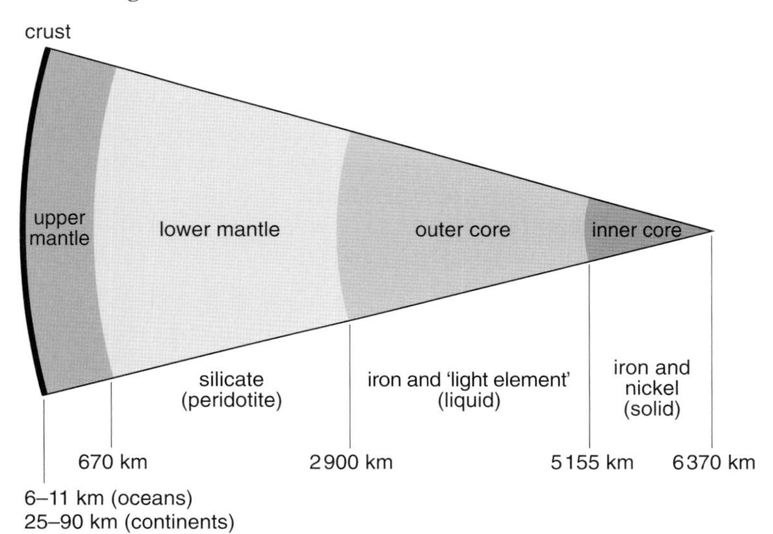

Figure 2.1 The compositional layers within the Earth, showing the seismically determined interfaces between each. The thickness of the crust has been exaggerated for clarity. The interface between the upper and lower mantle is thought to correspond to the transition of spinel to perovskite.

Table 2.2 Estimated global average composition of the mantle and the two types of crust, expressed as % oxide, showing also their approximate average densities at low pressure.

	Mantle	Oceanic crust	Continental crust
SiO_2	44.5	49	60
Al_2O_3	3	16	16
$FeO + Fe_2O_3$	8	9	7
MgO	38	9	3
CaO	3	11	6
Na_2O	0.4	3	3
K_2O	0.03	0.3	3
density/10^3 kg m^{-3}	3.3	3.0	2.7

Question 2.2 To which of the rock groups ultramafic, mafic, intermediate and felsic are the average compositions of (a) oceanic and (b) continental crust in Table 2.2 most similar?

It is important to remember that Table 2.2 shows *average* compositions only. The average composition of the oceanic crust is actually quite representative of the ocean floor, where it is rare to find extensive tracts of igneous rocks whose composition is other than mafic. However, continental crust is much more variable in composition and there are large volumes of continental crust, particularly in its upper part, whose composition is felsic rather than intermediate. There are also thick accumulations of sediment forming rock such as sandstone, mudstone and limestone, particularly on the continental shelves, whose compositions are similar to those listed in Table 7.2 of Block 2. The spatial variability of continental crust makes it particularly difficult to estimate its global average composition, so you may find values quoted elsewhere differing by several per cent from ours. However, there is no dispute that continental crust is broadly speaking intermediate in composition and thus significantly richer in SiO_2 than the mafic oceanic crust.

The crust–mantle distinction is significant because the crust (of either variety) is made from material that has been extracted from the mantle by melting processes (this is true even of the components of sedimentary rocks, if you trace them back far enough through the rock cycle). Mantle rocks are exposed at the surface only in exceptional circumstances, so in this Course we are chiefly concerned with crustal processes, particularly those involving the continental crust.

Where continental crust is joined to oceanic crust, there is a lateral transition from one type to the other (a **continental margin**) that may extend over tens of km. However, the junction between crust (whether oceanic or continental) and the underlying mantle can usually be located precisely because it is marked by a jump in the speed at which seismic waves are transmitted. This discontinuity in seismic speeds defines the crust–mantle boundary and is termed the Mohorovičić discontinuity (after its discoverer), usually abbreviated simply to the **Moho**.

The continental crust is complex and in its upper, more accessible, part shows little tendency towards well-defined internal layering (except in sedimentary rock), whereas oceanic crust has a simpler structure. Part of the reason for this contrast is that most of the continental crust is very old, going back several billion years, and has a complex history, whereas the oldest oceanic crust is only about 200 million years old. This does not mean there was no oceanic crust before that time; rather it is because the older that oceanic crust becomes, the greater its chances of being destroyed by the processes you will meet in Section 2.3. On the other hand, continental crust is virtually indestructible; it can be broken apart, rearranged, deformed and recycled via the rock cycle but it cannot be destroyed wholesale like oceanic crust.

Another difference between oceanic and continental crust is that oceanic crust is much thinner than continental crust. This is why virtually all oceanic crust lies well below sea-level whereas much of the continental crust is exposed on land. You will find out more about the relationship between crustal thickness and surface height shortly.

2.2.2 Mechanical layering

Although crust and mantle differ in origin and in the nature and variety of rock types that they contain, the crust does not form an independent layer in a mechanical sense. The Moho is a firmly welded interface, *not* a break permitting relative movement. However, if you are familiar with plate tectonics (which we will review in Section 2.3), you will be aware that the Earth's outer shell does move over the interior. In this context, the outer shell is defined by mechanical properties and consists of the crust and the top few tens of km of the mantle. The two are almost always inseparable, and form a discrete mechanical unit called the **lithosphere**. *Lithos* means 'rock' in Greek, and the term lithosphere was invented to describe the shell that behaves like rock in the familiar sense of being rigid. However, the application of external forces can cause parts of it to be deformed to produce structures of the kind you saw in Block 1, and that you will study in more detail in Sections 9–12.

Below the lithosphere, the rest of the mantle is also made of rock, in terms of its chemical and mineralogical composition. It, too, is solid rather than molten (in contrast to the outer core), but there is a crucial difference between it and the lithospheric part of the mantle. The prevailing pressure and temperature allow it to behave like a fluid when observed on time-scales of centuries or more. The forces acting to cause large-scale flow in the mantle below the lithosphere are chiefly buoyancy forces that drive convection (caused by heat generation, which you will read about in Section 2.4) and (at the very top) drag forces caused by sideways motion of the overlying lithosphere. This does *not* mean that this sub-lithospheric mantle is a liquid. Its rate of flow is only a few cm per year. Flow is achieved by migration of dislocations through crystals and adjustment of grain boundaries (Block 2 Activity 2.4), and also by movement between crystals lubricated by a thin film of melt at grain boundaries. The latter is particularly important in a 50- to 200-km-thick layer immediately below the lithosphere where the proportion of melt reaches a few per cent and is sufficient to reduce the speed of seismic waves by a few tenths of a km s^{-1}. This layer is variously known as the low-speed layer or **low-velocity zone**, and is the weakest part of the mantle.

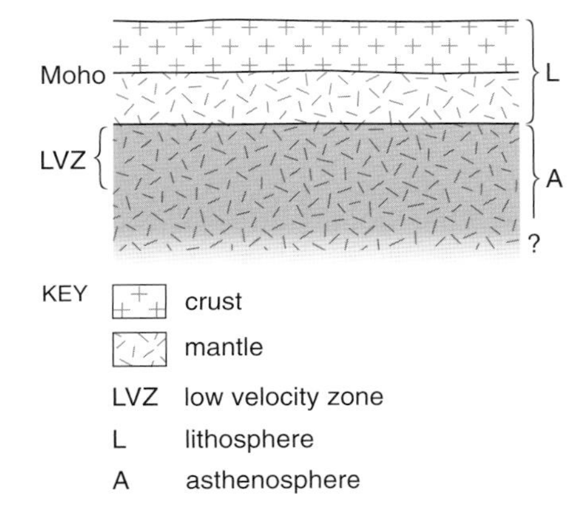

Figure 2.2 The division of the crust and upper mantle into two mechanical layers: the lithosphere and the asthenosphere.

The weak part of the mantle below the lithosphere is known as the **asthenosphere**, *astheno* coming from the Greek for 'weak'. Different people use the term in different ways. Some use it to refer only to the low-velocity zone, others use it to encompass the whole of the convecting part of the mantle (i.e., *all* the mantle between the lithosphere and the core), and a middle option confines it to that part of the mantle from the top of the low-velocity zone down to the 670 km interface between the upper and lower mantle. A precise definition is unnecessary for this Course, because we are not concerned with the deeper mantle. The mechanical layering of the crust and upper mantle is summarized in Figure 2.2.

2.2.3 SURFACE HEIGHTS: BUOYANCY AND ISOSTASY

The weakness of the asthenosphere allows lithospheric plates to slide around in the motion described as plate tectonics (Section 2.3). It is also ultimately responsible for the relative height differences between the surfaces of regions of continental and oceanic crust. Let's just think about continental crust for now.

The average thickness of continental crust is about 30 km, whereas the thickness of the continental crust in major mountain belts is about 90 km. However, when you consider that the summit of Mount Everest is less than 9 km above sea-level, it is clear that the heights of mountains are nowhere near as great as the extra thickness of crust underlying them.

❑ How can you explain this?

■ Most of the thickening of the crust that occurs in mountain belts must be achieved by lowering the *base* of the crust rather than merely raising up the *surface* of the crust.

The reason is that if the surface of the crust were raised without lowering its base, the crust would not be in equilibrium. Enormously high mountains could not be held up unless the underlying mantle were very strong (Figure 2.3). In practice, the weakness of the asthenosphere prevents this.

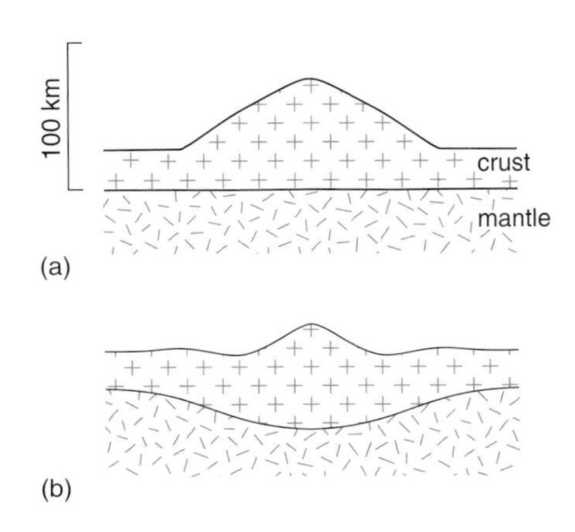

Figure 2.3 Cross-sections to show how unfeasibly high mountains created by unrealistic thickening at the top of the crust, (a), would subside to produce a balanced structure, (b).

The subsidence of the base of the crust between stages (a) and (b) of Figure 2.3 is achieved by bending of the underlying lithospheric part of the mantle. This is allowed *only* because this is in turn underlain by the much weaker asthenosphere which can flow out of the way. We do not show the base of the lithosphere in Figure 2.3, but you should not imagine it as running parallel to the crust–mantle boundary in (b). This is because asthenospheric conditions depend on pressure and temperature. Although the top of the asthenosphere will initially be depressed, when pressure and temperature have adjusted to the new situation (pressure instantaneously, temperature over millions of years), it may run approximately horizontally beneath the area. In other words, below the thickened part of the crust, the lowest part of what was formerly lithospheric mantle will become part of the asthenosphere.

However, the important point here is not what happens to the lithosphere–asthenosphere boundary. All you need to remember is that it is the weakness of the asthenosphere that would require the situation in (a) to subside to the situation in (b). But why should (a) subside to (b)?

❑ Can you think of a simple reason?

■ It is simply because of the extra weight of the thickened part of the crust.

Although it is only a sketch, Figure 2.3b is meant to show a situation in balance, where the weight of each part of the crust is supported because it 'floats' on the mantle beneath it. To see how this works, refer now to Figure 2.4, and recall that the crust is less dense than the mantle.

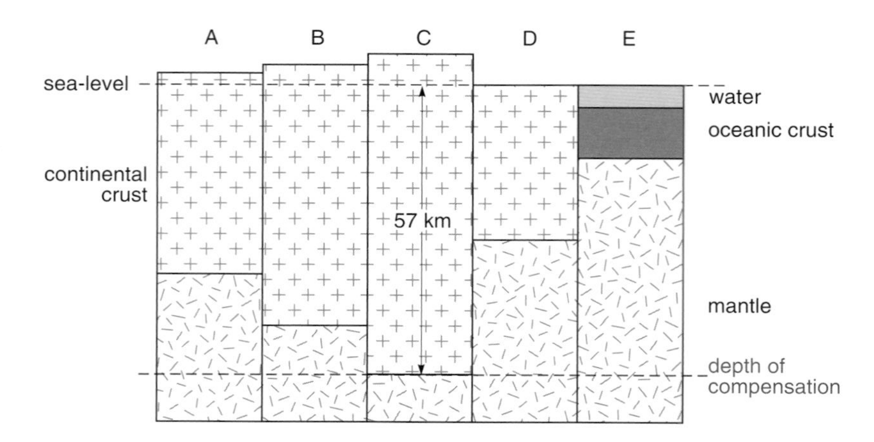

Figure 2.4 Cross-section dividing the crust into blocks A–E resting in isostatic equilibrium upon the mantle. This shows how mountains of modest height (6 km above sea-level in column C in this example) are underlain by a deep crustal root whose buoyancy supports them. The densities of water, continental crust, oceanic crust and mantle are 1×10^3, 2.7×10^3, 3×10^3 and 3.3×10^3 kg m^{-3} respectively.

In the situation shown, the pressure at the base of each column (crust plus mantle, and including water in column E) is the same. This means that no strength at all is required to keep the high blocks up or to hold the low blocks down. The blocks of crust 'floating' on the mantle can be likened to icebergs floating in water (except that the top of the mantle is too strong to flow, and it is only the weakness of the underlying asthenosphere that allows equilibrium to be achieved). The tops of the thicker crustal blocks are higher, but their bases go down much further. The principle whereby blocks of crust sit at their neutrally buoyant level is called **isostasy**, and the situation shown in Figure 2.4 is described as one of **isostatic equilibrium**.

Activity 2.1

To discover for yourself how this equilibrium happens, work through Activity 2.1. Doing so will also help consolidate your skills in using and rearranging equations.

Of course, the crust is not really divided into independent columns nor do the surface and Moho rise and fall in discrete steps as implied in Figure 2.4. However, the Activity you have worked through is a valid representation of the principle of isostasy. The Earth's crust is close to isostatic equilibrium throughout the globe, and every mountain belt is underlain by a thick 'root' or 'keel' like that shown in Figures 2.3b and 2.4.

Question 2.3 With reference to Figure 2.4, suggest two reasons why the top of the oceanic crust tends to be below sea-level.

2.3 PLATE TECTONICS

The significance of isostasy will re-emerge when you study what happens during collisions between continents. The main point for now is that the lithosphere can flex and move sideways because of the weakness of the underlying asthenosphere. The thickness of the lithosphere varies from place to place, and is greater beneath continents than beneath oceans. In this Course, we will define the lithosphere as the rigid layer that takes part in plate tectonic movements. This is as thin as 40 km beneath the oceans but may exceed 100 km beneath parts of the continents. (Elsewhere you may come upon alternative definitions of the lithosphere, based on thermal, seismic, or elastic properties, which may have slightly different thicknesses.)

The lithosphere is not an intact shell; rather it is divided into several large pieces called **plates** (Figure 2.5) consisting of crust and the immediately underlying lithospheric part of the mantle. Most plates contain both continental and oceanic crust. Each plate is moving relative to its neighbours, at typical speeds of a few cm yr^{-1}. This happens in such a way that gaps do not open up between plates, and in this respect the Earth's lithospheric shell is unbroken.

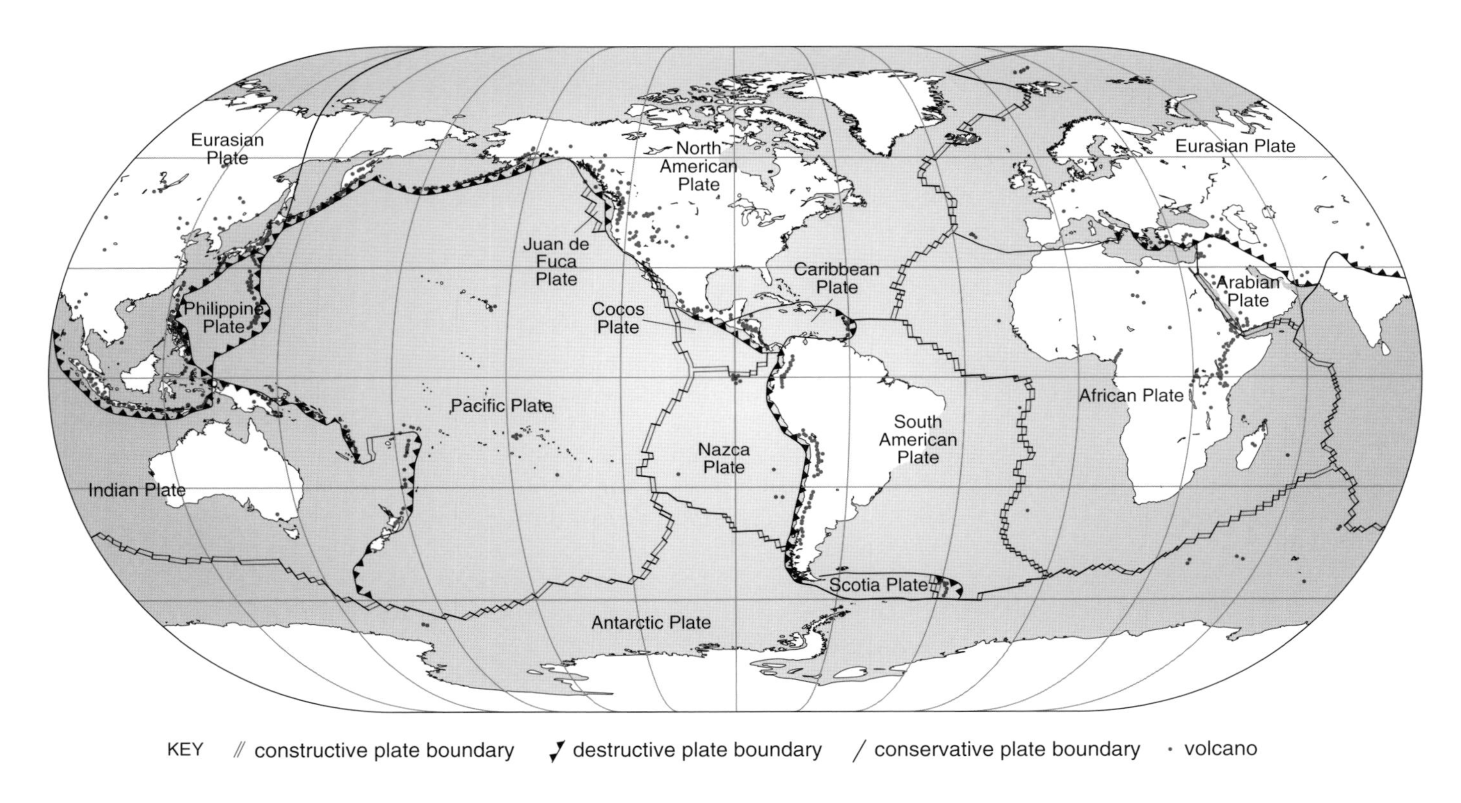

Figure 2.5 A map of the Earth's tectonic plates. The three types of **plate boundary** identified in the key are discussed in the text. The triangles along the destructive (convergent) plate boundaries are on the plate that is going *over* its neighbour (but note that they do not point in the direction towards which that plate is moving).

We will review plate tectonic processes very briefly by considering boundaries between adjacent plates in three circumstances, i.e. where the relative motion is convergent, divergent, and parallel.

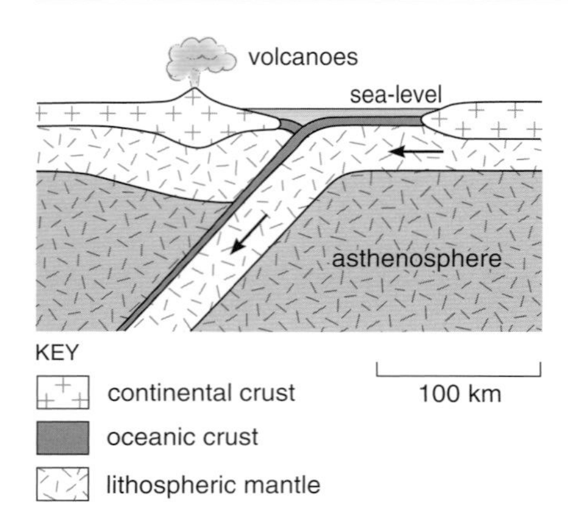

Figure 2.6 A cross-section through a destructive plate boundary. The angle of descent shown here is fairly typical, but extreme cases dip as gently as 15° or as steeply as 80°.

2.3.1 PLATES CONVERGING

Where two plates are moving towards each other, one of them must dive down below the other. But which one?

❑ Suppose the top of the colliding edge of one plate consists of continental crust, whereas the top of the colliding edge of the other plate is oceanic crust. Bearing in mind the relative densities of the two types of crust, which plate would it be easiest to force downwards?

■ Oceanic crust is denser than continental, so the edge of the plate carrying oceanic crust would tend to go below the continental plate edge.

This situation is illustrated in Figure 2.6, which shows a **destructive plate boundary***, so called because the downgoing plate is destroyed. The act of one plate diving below another is termed **subduction** and the relevant part of a destructive plate boundary is described as a **subduction zone**.

❑ What are two reasons why the subducting part of the plate should be destroyed?

■ One is that as it gets pushed deeper it will encounter progressively higher temperatures, and begin to melt. Another is that as it encounters higher pressures, phase transitions will cause the minerals within it to change into denser forms. Both processes tend to destroy the integrity of the plate.

Where the two plates grind together, earthquakes are triggered. By determining the locations of these earthquakes in three dimensions, the downgoing plate can usually be traced to depths of several hundred km. Destructive plate boundaries are also characterized by volcanoes on the overriding plate. As you will see later in this Block, this is a consequence of melts escaping upwards both from the subducting plate and from the base of the overriding plate. The Andes volcanoes of South America are a classic example.

When the colliding edges of both plates carry oceanic crust, it is not so obvious which one will be destroyed. Generally speaking, the plate with the oldest, and therefore coldest and densest, oceanic crust is the one to be subducted. When a subduction zone occurs within an ocean, volcanoes appear near the leading edge of the overriding plate, for the same reasons as in an ocean–continent collision. The volcanic **island arcs** of the western Pacific Ocean and the Caribbean are good examples. If volcanism (and associated intrusion of igneous rocks at depth) is prolonged, the belt of affected crust can be so thoroughly altered that it becomes more continental than oceanic in character, and in fact this has happened in the case of Japan.

Subduction-related volcanoes occur only on the overriding plate, and tend to be found about 100 km above the subduction zone. Where the two plates meet, the downward bend of the subducting plate means that the sea-floor is particularly deep. The surface expression of a subduction zone is therefore a **trench** reaching typically about 8 km below sea-level.

When two plates are converging, the main driving force is believed to be the weight of the relatively dense oceanic lithosphere pulling it below the less dense continental lithosphere (or the island arc) of the overriding plate. Once this situation is established, there appears to be nothing that can stop it continuing until something really drastic happens. In Figure 2.6, the subducting plate is shown as part oceanic and part continental, with the oceanic part at its leading edge being subducted. What will happen if we wind the clock forward?

* In S103 *Discovering Science*, this is called a convergent plate boundary.

Question 2.4 In Figure 2.6, there is about 80 km of oceanic lithosphere left to be subducted. If the plate on the right is moving towards the plate on the left at a rate of 1 cm yr^{-1}, how long will it be before all the oceanic lithosphere has been subducted?

The situation after this time is shown in Figure 2.7a. The weight of the subducting oceanic part of the plate will continue to drive the convergent motion, but now it has to pull down continental crust. This is less dense and thus more buoyant than oceanic crust, and cannot be pulled down far. In such a continent–continent collision, the crust is thickened by a combination of compressional deformation of the leading edge of the overriding continent and because part of the continental crust on the other plate begins to be subducted. Eventually the collision locks up, and the oceanic part of the subducting slab may break free, as indicated in Figure 2.7b. The result at the surface is a **suture zone** marked by a mountain belt (an isostatic outcome of crustal thickening) and often a few slivers of oceanic crust and upper mantle that have avoided subduction. You will study many consequences of continent–continent collision later in this Block.

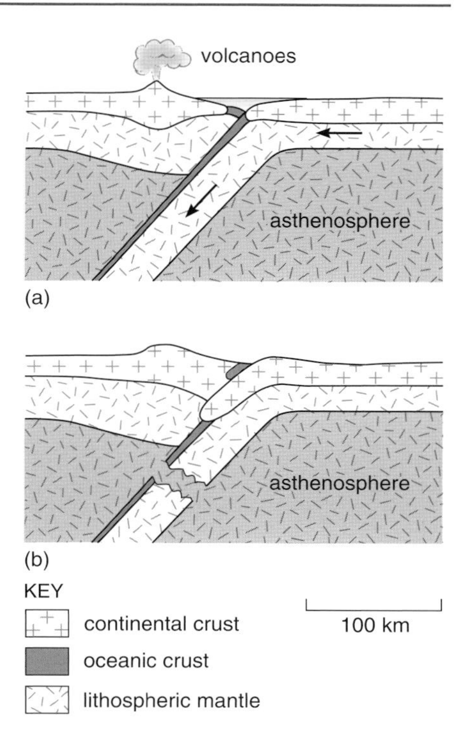

Figure 2.7 (a) Cross-section to show the outcome of continuing subduction for several tens of millions of years beyond the stage shown in Figure 2.6. (b) A few tens of millions of years later still, a continent–continent collision has forced subduction to cease, and the slab of subducted oceanic lithosphere may break free, as illustrated here.

2.3.2 PLATES MOVING APART

When two plates diverge, new oceanic lithosphere is created between them in a process aptly named **sea-floor spreading**. This stops a gap opening up, as shown in Figure 2.8. Such a plate boundary is described as a **constructive plate boundary***.

In sea-floor spreading, the divergent plate motion draws the asthenosphere upwards in a linear belt of upwelling below the plate boundary. As it rises the asthenosphere begins to melt, releasing mafic magma that rises even faster and creates new oceanic crust. The upwelled asthenospheric mantle cools as it approaches the surface and adheres at roughly equal rates to the plates on either side of the boundary, becoming part of their lithosphere. Both the crustal and mantle parts of the newly formed lithosphere are still relatively warm, making it slightly less dense and thus more buoyant than the older and colder lithosphere that has been displaced further from the boundary. Constructive plate boundaries are therefore marked by ridges on the ocean floor. Typically, the crest of such a ridge is 2–3 km below sea-level, but as it moves away to either side the ocean floor cools and subsides isostatically to an average depth of 4–5 km. One of the best-known such ridges is the Mid-Atlantic Ridge that snakes its way from north to south through the Atlantic Ocean (Figure 2.5) and permits the present day east–west divergence of North and South America from Europe and Africa, at a rate of a few cm yr^{-1}. Although not all of them are midway across an ocean, constructive plate boundaries are often referred to as **mid-ocean ridges**.

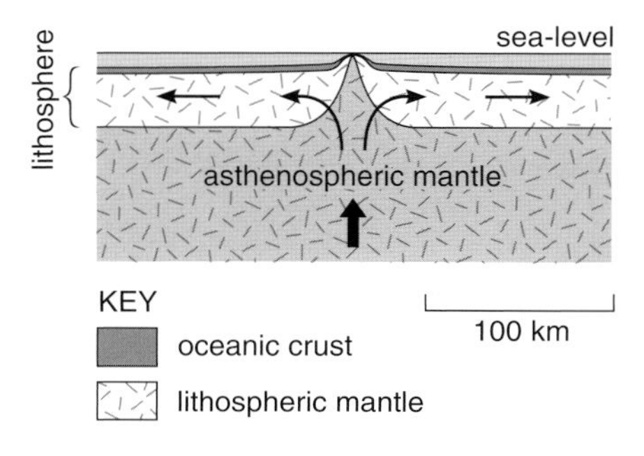

Figure 2.8 Cross-section through a constructive plate boundary, drawn to the same scale as Figures 2.6 and 2.7.

Divergent motion cannot occur with oceanic lithosphere on one side and continental lithosphere on the other, at least not for any length of time, because divergent motion creates oceanic lithosphere. This explains why constructive plate boundaries are found today only within oceans (Figure 2.5). However, it sometimes happens that divergent motion is initiated within a continent. What happens in such a situation is shown in Figure 2.9 (overleaf).

The whole region in Figure 2.9a belongs to a single plate, but by the stage shown in Figure 2.9c the two halves are separated by a constructive plate boundary like that in Figure 2.8 and so belong to two separate plates. What was initially a single continent has rifted into two parts, each joined to new oceanic lithosphere at newly formed continental margins. The time interval

* In S103 *Discovering Science*, this was called a divergent plate boundary.

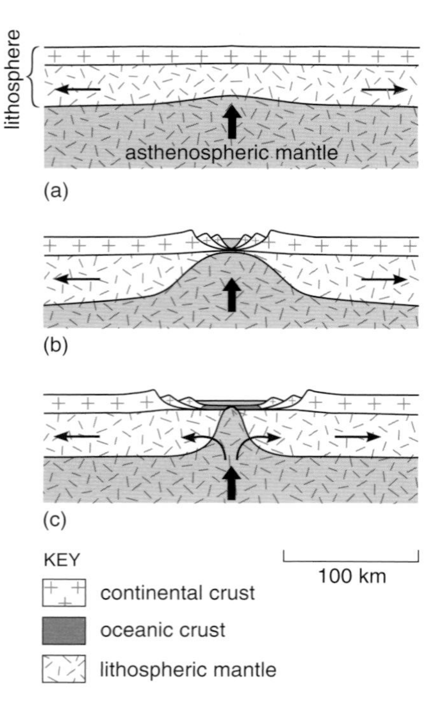

Figure 2.9 Time-series of cross-sections at the same scale as Figures 2.6 to 2.8 illustrating divergent motion within a continent. In (a) and (b), upwelling within the asthenosphere causes heating and thinning of the continental lithosphere. By stage (c), the continent has rifted apart and a new ocean (like the Red Sea, between Arabia and Africa) is beginning to open between them.

between stages (a) and (c) is of the order of 100 million years. The faults shown in the process of formation in Figure 2.9b are normal faults characteristic of an extensional tectonic regime, and are common in continental margins formed by rifting. You will learn more about this type of situation in Section 12.2.1.

2.3.3 Oceans opening and closing

Elements of the processes shown in Figures 2.6–2.9 are combined in Figure 2.10 to illustrate the life cycle of an ocean. In Figure 2.10a, a young ocean is opening. In stage (b), a subduction zone has formed near one side of the ocean. Whether or not the ocean continues to widen depends on the relative rates of subduction and sea-floor spreading. In this example, the two rates are more or less matched so that the ocean stays roughly the same width between stages (b) and (c). However, by the time stage (c) is reached, the constructive plate boundary has built itself out close to the subduction zone and will soon be lost. This has happened by stage (d); the ocean basin now lacks a constructive plate boundary, and will eventually be destroyed in a continent–continent collision. The whole cycle is reckoned to take on average about 400–500 million years.

The present northern Atlantic Ocean is at the equivalent of stage (a), having a mid-ocean ridge but no destructive plate boundaries. The central Atlantic Ocean may be likened to stage (b), with subduction occurring beneath the Caribbean. The Pacific Ocean is most similar to stage (c). Its constructive plate boundary is displaced to the east side of the ocean, as you can see in Figure 2.5, but the situation is complicated because the Pacific Plate is being subducted on the opposite (western) side of the ocean too.

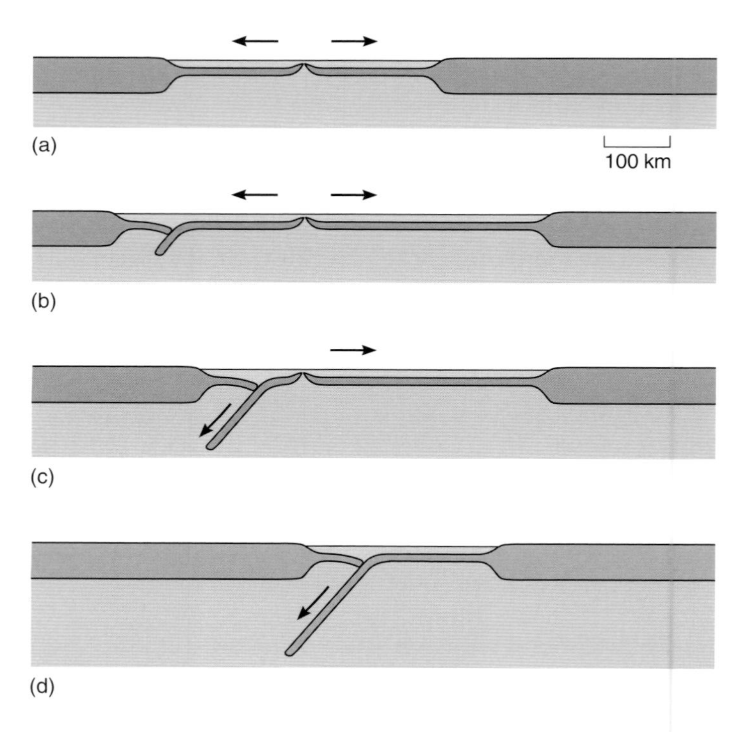

Figure 2.10 Time series of cross-sections to show the opening and closing of an ocean. The scale is reduced compared to that of Figures 2.6 to 2.9, and crust and mantle are not distinguished within the lithosphere. See text for discussion.

2.3.4 PLATES MOVING SIDEWAYS

We may have given the impression that the relative motion between adjacent plates is always perpendicular to the boundary between plates. This is generally true in the case of constructive plate boundaries. In contrast, collision can be at any angle. Oblique collision need not concern us for now, and indeed in cross-section it would look indistinguishable from the situation shown in Figures 2.6 and 2.7.

However, there is a third type of plate boundary that we have yet to consider, which is where adjacent plates slide past each other. In such a situation there is neither creation nor destruction of lithosphere, so it is described as a **conservative plate boundary**. These are most common in the oceans, where constructive plate boundaries are offset by tens or hundreds of km along conservative plate boundaries in the form of **transform faults** (Figure 2.11).

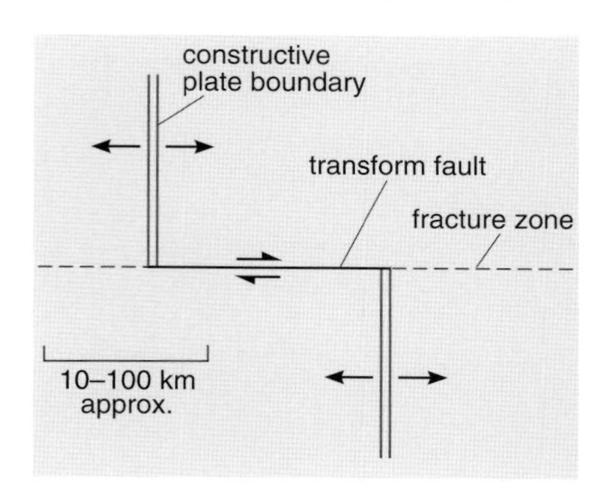

Figure 2.11 Sketch map showing how a transform fault offsets a constructive plate boundary. A fracture zone is the apparent continuation of a transform fault beyond the overlap, usually manifested by a scarp; however, this is not a plate boundary and there is little or no motion across it.

Conservative plate boundaries can occur within continental crust, too. The San Andreas fault of California is the most famous example, and it is worth looking back at Figure 2.5 at this stage. This depicts the San Andreas fault rather schematically. It links the constructive plate boundary between the Pacific Plate and the Cocos Plate to the constructive plate boundary further north between the Pacific Plate and the Juan de Fuca Plate. Looking more generally at this map, you can see (allowing for the limitations of scale and the distortion caused by portraying the Earth's surface on a flat page) that conservative plate boundaries tend to be at right angles to whatever constructive plate boundary they offset, but that destructive plate boundaries can meet other kinds of plate boundary at any angle. Note also that destructive plate boundaries are curved, particularly where they occur between oceanic parts of plates. An arcuate pattern is especially clear in the northern and western Pacific.

❑ What topographic features would you expect to find on the inner (convex) side of each arc?

■ The convex side of the arc is on the overriding plate. We would expect volcanoes to develop above the subduction zone, which would therefore form volcanic islands.

These, of course, are the island arcs that you met in Section 2.3.1.

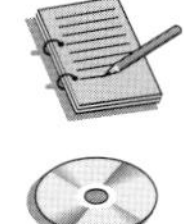

Activity 2.2

We hope that by now you have a clear mental picture of the relative movements between plates that plate tectonics involves, and of how plates are created and destroyed. However, if you are still puzzled, the animations that we have provided for Activity 2.2 should help. There is less than a minute of playing time in total, and it should not take you more than about 10 minutes to complete. Do this *now* if you feel you need help right away; otherwise, do it next time it is convenient to use the DVD on your computer.

2.4 THE EARTH'S HEAT

We have already told you that the Earth's interior is hot, and that the sub-lithospheric mantle is convecting. This convection transfers heat outwards from the interior towards the base of the lithosphere. Most of the heat reaching the Earth's surface does so through the creation of new, hot lithosphere at constructive plate boundaries, counterbalanced by the subduction of old, cold lithosphere at destructive plate boundaries, which cools the interior. However, some heat escapes through the lithosphere by conduction and also by upward movement of bodies of magma. The total average global rate of heat loss is about 4×10^{13} W*. This works out at an average heat flow of about 0.08 W m^{-2}, though it varies considerably from place to place.

We have not yet mentioned the origin of the Earth's heat. This is relevant for understanding many of the processes that you will study later in this Block. Between about 20% and 50% is reckoned to be primordial heat inherited from the time of the Earth's formation, and the remainder represents heat that is still being generated. The only important heat source today involves the decay of radioactive elements. You should remember from Video Band 1 that measuring the products of radioactive decay provides the basis of radiometric dating. All radioactive decay produces heat, which is described as **radiogenic heating**, but there are only three elements today whose decay produces significant amounts of heat. These are uranium, thorium and potassium.

Thorium (Th) is the simplest case, because in the Earth it consists almost entirely of a single isotope, ^{232}Th**, which decays with a half-life of 13.9 billion years. This is too slow a rate to be of much use in radiometric dating but nevertheless this isotope acts as a significant heat source. Natural uranium (U) occurs as a mixture of two isotopes, ^{235}U and ^{238}U, whose half-lives are suitable to be of use in radiometric dating. Because ^{235}U has a shorter half-life than ^{238}U, the proportion of ^{235}U relative to ^{238}U has been declining ever since the Earth's origin and today only 0.7% of uranium is ^{235}U. Most potassium consists of the stable isotope ^{39}K, but 0.01% of it today is the radioactive isotope ^{40}K, whose decay to argon is much used in radiometric dating.

Table 2.3 lists the relevant isotopes of these **heat-producing elements** and shows the rate of heat generation per kg of each isotope.

Table 2.3 Half-lives and rates of heat generation for isotopes of the important heat-producing elements in the Earth today.

	Half-life / 10^9 yr	Heat generation / 10^{-5} W kg^{-1}
^{40}K	1.30	2.8
^{232}Th	13.9	2.6
^{235}U	0.71	56
^{238}U	4.50	9.6

Table 2.3 shows that ^{40}K and ^{232}Th produce roughly similar amounts of heat per kg of isotope. However, given that only 0.01% of all K is ^{40}K, you might be surprised that ^{40}K is an important heat source. The explanation is that K is several orders of magnitude more abundant in the Earth than either Th or U, so even though only a tiny amount of K is ^{40}K there is still enough ^{40}K to be significant.

The absolute and relative abundances of the heat-producing elements in different kinds of rock are crucial in controlling where in the crust and mantle most of the heat is produced. You already know from Table 2.2 that K is about ten times more abundant in continental crust than in oceanic crust, which is itself ten times richer in K than the mantle. Therefore, each kg of continental crust must be heated by ^{40}K decay at something like ten times the rate per kg of oceanic crust and at a hundred times the rate per kg of mantle. To see if this pattern is followed for all heat-producing elements, look now at Table 2.4 and attempt Question 2.5.

* W = watt. This is the SI unit of power. 1 watt = 1 joule per second (1 W = 1 J s^{-1}).

** This means the element thorium with 232 heavy particles (protons + neutrons) in its nucleus. It is usually referred to in speech as 'thorium-two-three-two'.

Table 2.4 Concentrations of the main heat-producing isotopes (in parts per million, ppm) and the resulting rate of heat generation per kg of rock. The total heat generation in granite (final column) is to be calculated in Question 2.5

Rock type	K (total)	^{40}K (0.01% of all K)	^{232}Th (all Th is ^{232}Th)	U (total)	^{235}U (0.7% of all U)	^{238}U (99.3% of all U)	Total heat generation/ 10^{-10} W kg^{-1}
granite	4×10^4	4	20	4	3×10^{-2}	3.97	
basalt	4×10^3	0.4	2	0.5	3×10^{-3}	0.497	1.1
peridotite	4×10^2	0.04	0.06	0.02	1.4×10^{-4}	0.02	0.047

Question 2.5

(a) Use the data on heat generation per kg of isotope given in Table 2.3 to verify the total heat generation in basalt and peridotite in Table 2.4.

(b) Calculate for yourself the total heat generation in granite using the same method. Enter your value in the relevant space in Table 2.4.

We hope you noticed while tackling Question 2.5 that the abundances of all three heat-producing elements, and therefore the rate of radiogenic heat production, are much greater in granite (which is a felsic rock type) than in basalt (mafic), and that they are even less in peridotite (ultramafic). It takes no great leap of the imagination to deduce that the elemental abundances and rate of radiogenic heat production in intermediate rocks fall somewhere between the values for granite and basalt.

In global terms, taking into account the fact that the mass of the mantle is two orders of magnitude greater than that of the crust, the total radiogenic heat generated in the whole mantle works out at approximately equal to that generated in the whole crust. Radiogenic heating is greater in continental crust than in oceanic crust, and is particularly great in the upper part of the continental crust where most of the felsic rocks are concentrated. This has an important influence on the result of continent–continent collision. As you saw in Section 2.3.1, such an event leads to major thickening of the crust, so the total rate of heat generation underlying each square metre of surface is significantly greater in collision zones. You will explore some of the consequences of this in Sections 7.5 and 8.4.

2.5 SUMMARY OF SECTION 2

- The Earth is layered compositionally into inner core, outer core, mantle and crust. There are two kinds of crust: oceanic and continental.
- The outer part of the Earth can be divided into two mechanical layers: the rigid lithosphere, consisting of the crust and the uppermost mantle, and the weak asthenosphere underlying it.
- Below the lithosphere, the mantle convects to transport heat outwards. However, it is solid.
- The low velocity zone immediately below the lithosphere is the weakest part of the asthenosphere. The weakness of this layer enables (i) the crust to reach isostatic equilibrium, and (ii) lithospheric plates to slide over the asthenosphere.
- Oceanic lithosphere is created at constructive (divergent) plate boundaries and destroyed at destructive (convergent) plate boundaries. Most of the continental crust is much older than any surviving oceanic crust.
- Most of the heat escaping from the Earth's interior is generated by decay of radioactive isotopes of U, Th and K, and these are particularly concentrated in the continental crust.

2.6 Objectives for Section 2

Now you have completed this Section, you should be able to:

2.1 Describe the Earth's internal structure, and indicate how this is known.

2.2 Describe the rudiments of plate tectonics, giving examples of different kinds of plate boundary.

2.3 Describe how heat is generated within the Earth and use relevant data to calculate the rate of radiogenic heating in different rock types.

Now try the following questions to test your understanding of Section 2.

Question 2.6

(a) How do each of the following provide information on the Earth's interior: (i) the ratio between its mass and volume, (ii) the way in which seismic waves pass through it, (iii) its magnetic field?

(b) Which of these would you say is the most important?

Question 2.7

(a) Imagine that the crust at the top of column C in Figure 2.4 is subjected to erosion. In what way will the depth of the base of the crust in column C change as the top of column C is eroded, and why?

(b) Suppose the sediment derived by erosion from the top of column C is deposited on the top of column D. What will happen to the crust in column D?

(c) Where is the actual flow taking place that allows the base of the crust to change its depth?

Question 2.8 Briefly describe three independent lines of evidence that indicate the presence of a weak asthenosphere immediately below the lithosphere.

Question 2.9 Is the following statement true or false? 'The whole of the Atlantic Ocean is underlain by a single tectonic plate.' Give your reasons.

Question 2.10 Where would you expect to find most heat per m^2 reaching the Earth's surface by conduction, and why? (a) On old ocean floor; (b) on a region of continental crust slightly above sea-level; (c) in a mountain belt produced by continent–continent collision.

3 Igneous rocks and the rock cycle

Let's pause now to see how what we have covered in this Block so far bears on the rock cycle that you met in Section 9 of Block 2, and in particular how igneous rocks fit in. Oceanic crust is created by igneous processes at constructive plate boundaries, and is recycled back into the mantle at destructive plate boundaries a few hundred million years later. There is almost a closed loop here, but some of the magma contributing to volcanoes above subduction zones derives from the downgoing oceanic plate, so Andes-type settings provide a mechanism for elements to pass from the mantle to the continental crust via the oceanic crust.

Volcanic and intrusive igneous rocks in highland regions such as volcanic arcs and continental collision zones are prone to erosion. As noted in the answer to

Question 2.7a, erosion of the surface of thickened crust promotes further isostatic uplift. This will continue until the crust has been reduced to normal thickness and its surface is at the average level. Sediment derived by erosion of thickened crust is deposited elsewhere on the crust and some must find its way into ocean trenches at subduction zones. Most of this usually gets scraped onto the front of the overriding plate (eventually forming part of the thickened crust resulting from collision) but some may be carried down into the subduction zone on top of the downgoing slab, where it can become melted and contribute to the formation of a new generation of igneous rocks in an island arc or Andean volcanic chain.

Activity 3.1

You should now do Activity 3.1, which is intended to consolidate your grasp of where igneous rock types are characteristically found.

3.1 Objectives for Section 3

Now you have completed this Section, you should be able to:

3.1 Give examples of how rock formed as a consequence of plate tectonics can become recycled.

3.2 State the igneous rock types most likely to form in each plate tectonic setting, but be aware of possible exceptions.

Now try the following question to test your understanding of Sections 2 and 3.

Question 3.1 With reference to Figure 2.5, describe how oceanic crust created today at the constructive plate boundary between the Pacific Plate and the Nazca Plate could eventually provide material that adds to the continental crust of South America.

4 Melting rocks

4.1 What causes melting?

Now is the time to consider how magma is produced. What actually causes rock to melt? You might think the answer is obvious, and that magma must be a result of heating. In fact, although sufficient heating of a volume of rock *will* yield magma, heating is not the only, or even main, cause of melting in the crust and upper mantle.

We will consider other causes of melting shortly, but first let's think about what happens during melting. Recall from Section 6.2 of Block 2 that when a magma solidifies as a result of cooling, different minerals begin to crystallize at different temperatures. In melting, the converse is the case (irrespective of whether melting is caused by heating or some other factor). In any particular rock consisting of more than one mineral, all minerals *begin* to melt at the same temperature but as melting proceeds each mineral contributes to the melt at a different rate until it is entirely used up. Minerals disappear sequentially in the reverse of the crystallization sequence shown in Plate 6.5 of Block 2. Usually, the process stops while there is at least one mineral surviving, so that the rock is still partly crystalline. This behaviour is termed **partial melting**.

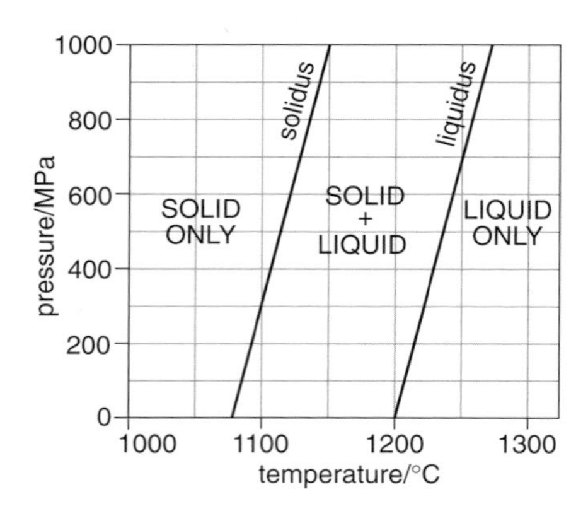

Figure 4.1 Phase diagram showing the solidus and liquidus for a typical anhydrous mafic rock. Pressure is shown in megapascals (MPa = 10^6 Pa). The **pascal (Pa)** is the SI unit for pressure (1 Pa = 1 N m^{-2} = 1 kg m^{-1} s^{-2}). Elsewhere you may find kilobars (kbar) used to denote geological pressures: 1 kbar (a thousand times atmospheric pressure at sea-level) is 100 MPa.

For a rock of a particular composition, we can find a temperature at which melting will begin and a higher temperature at which melting will be complete. These two temperatures vary according to pressure, and it is conventional to show this relationship on a graph of pressure (P) against temperature (T) with a line joining all points where melting begins (called the **solidus**) and another line joining all the points where melting is complete (called the **liquidus**). Such a P–T diagram is plotted for mafic rock in Figure 4.1. This is a **phase diagram** analogous to those you met in Section 2 of Block 2, with solidus and liquidus bounding areas in which the material is in different states.

Figure 4.1 is plotted for anhydrous conditions (i.e. no water present). As you will see shortly, the presence of water has a major influence. In the meantime, Question 4.1 tests whether you can read and understand the information given by this diagram.

Question 4.1

(a) Decide whether material of mafic composition consists of solid, liquid, or both under the following P–T conditions: (i) atmospheric pressure (0.1 MPa), 1100 °C, (ii) 1000 MPa, 1100 °C, (iii) 400 MPa, 1150 °C, (iv) 200 MPa, 1250 °C.

(b) If we have an anhydrous piece of basalt and heat it up keeping the pressure constant at 0.1 MPa, at what temperature does it (i) begin to melt, (ii) become completely molten?

As the temperature increased from 1080 °C to 1200 °C (in the answer to Question 4.1b), the sample would consist of a decreasing proportion of solid and increasing proportion of liquid. Just to the right of the solidus it would be mostly solid rock containing a tiny amount of melt, and just before it reached the liquidus it would be mostly melt (magma) with a few crystals dispersed in it. Mafic rock consists of olivine, pyroxene and plagioclase feldspar. Using Block 2 Plate 6.5 in reverse, you can deduce that the last crystals to survive would be olivine.

Pressure has an influence on partial melting too: try Question 4.2.

Question 4.2 With reference to Figure 4.1, describe what would happen if we took an anhydrous piece of basalt at 1000 MPa and 1100 °C and gradually decreased the pressure to 0.1 MPa without changing the temperature.

Thus, partial melting can be brought about by changing (in this example, decreasing) pressure *without* changing the temperature, and so a source of heat is not necessary. This is called **decompression melting**. It is also important to realize that when a sample is partially molten the composition of the liquid is *different* from the average composition of the remaining solid (because some minerals have contributed disproportionately to the melt), and that both are different to the average composition of the starting material. However, the average composition of the melt plus remaining solid must always be identical to the average composition of the starting material.

❑ We have seen that in the case of a partially melting basalt consisting of olivine, pyroxene and plagioclase, some of the olivine would survive as crystals after everything else had melted. If the melt were to be squeezed out, leaving these crystals behind, what name would you give to the rock formed by these crystals? (See Block 2 Plate 6.12.)

■ A rock consisting of olivine and little else would be ultramafic in composition and would be described as a peridotite.

Partial melting of mafic material could therefore leave a solid residue of ultramafic composition. On the other hand, the very first melt to form would be intermediate in composition because most of its constituents would be from pyroxene and the least calcium-rich plagioclase. Whether or not *all* the plagioclase and pyroxene had melted would depend both on their compositions and on the final *P–T* conditions. Broadly speaking, the closer the final conditions to the solidus, the richer in silica the melt but the smaller the compositional change in the solid residue. Conversely, if the final *P–T* conditions lay only just to the left of the liquidus, the less solid residue there would be and the more ultramafic its composition, whereas the melt would be only slightly richer in silica than the starting material.

In this Course, we are more concerned with the melt rather than the solid residue, so we can generalize the above as follows:

> When conditions are such as to end up with partial rather than total melting, the melt is richer in silica than the starting material. It is common for partial melting of ultramafic starting material to yield a mafic melt, for partial melting of mafic starting material to yield an intermediate melt, and for partial melting of an intermediate rock to yield a felsic melt.

There is a third factor controlling melting that we have yet to consider. Figure 4.1 is plotted for anhydrous conditions. By this we mean that there is no water dissolved in the magma, the solid contains no hydrous minerals, and water does not permeate the rock. As soon as water is added to the system, the whole situation changes. Water has the effect of lowering the melting temperatures, so that both the solidus and liquidus are displaced to the left on a *P–T* phase diagram. Figure 4.2 shows a solidus and liquidus plotted for water-saturated conditions, which are defined as those in which the liquid (magma) phase contains as much dissolved water as it can hold. Typically, this is a few percent (by mass). If there is more water present than the magma can hold, the water will exsolve, that is to say come out of solution and form a separate phase (usually in the form of vapour bubbles). This phenomenon is described as **exsolution**.

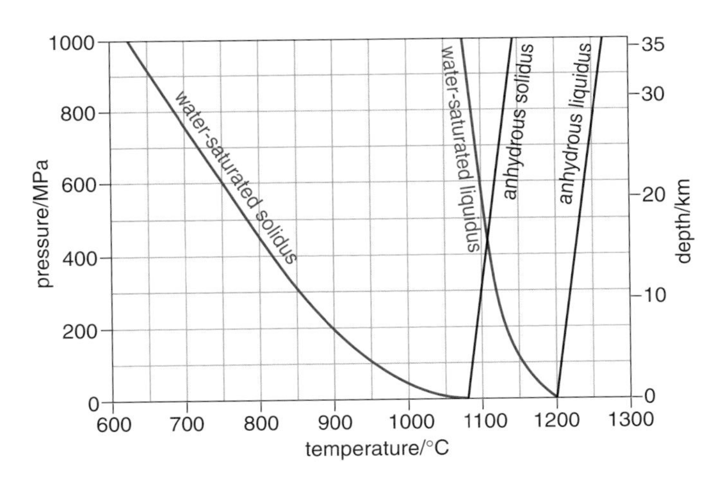

Figure 4.2 Phase diagram for mafic rock showing solidus and liquidus for water-saturated conditions, in addition to the anhydrous lines shown in Figure 4.1. In hydrous conditions but with insufficient water for saturation, the phase boundaries would lie between the anhydrous and water-saturated extremes. The vertical scale on the right indicates the approximate depth in continental crust corresponding to the pressures on the opposite scale.

On Figure 4.2, the water-saturated phase boundaries occur at lower temperatures than the anhydrous phase boundaries except at vanishingly small pressure. This is because if there is no confining pressure on the magma, virtually all the water will come out of solution and form bubbles, thereby making the melt anhydrous. To explore an important consequence of hydration, you should now try Question 4.3.

> **Question 4.3** What would happen if you took anhydrous mafic material at 1000 MPa and 1100 °C and (without changing *P* or *T*) added sufficient water to make conditions water-saturated?

Thus, melting can occur simply as a result of the addition of water. In fact anything else that can be held in magma as a dissolved vapour has a similar effect. Such substances are referred to as **volatiles**; water (H_2O) is the most abundant but carbon dioxide (CO_2), sulfur dioxide (SO_2), and fluorine (F_2) are also common. Water saturation has an even more extreme effect on the melting behaviour of felsic rock, as you can see in Figure 4.3, which we will be using several times later in the Block.

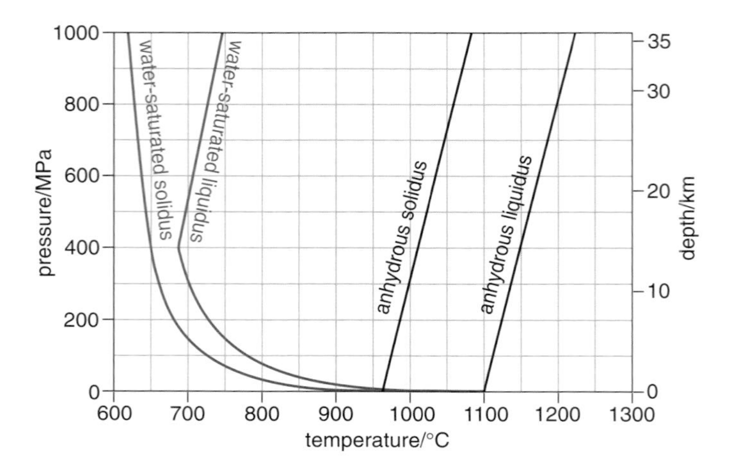

Figure 4.3 Anhydrous and water-saturated phase boundaries for felsic rock. Compare with Figure 4.2. Note that in both cases the water-saturated phase boundaries meet the equivalent anhydrous phase boundaries at zero pressure.

4.2 WHAT IS A MAGMA?

At this point, we need to say a little about what silicate magmas consist of. Section 4 of Block 2 showed how most rock-forming minerals are based on the SiO_4 tetrahedron. The bonds between Si and O within this tetrahedron are stronger than other bonds in silicate minerals. When a mineral melts it is the weaker bonds that break. The SiO_4 tetrahedra tend to remain intact but metallic elements form positively charged ions such as Ca^{2+}, Mg^{2+}, Fe^{2+}, Na^+ and K^+. In an ultramafic magma, the SiO_4 tetrahedra occur mostly as isolated negatively charged SiO_4^{4-} ions. However, in a magma that is mafic (and therefore richer in silica than an ultramafic magma) some of the SiO_4 tetrahedra are linked together (polymerized) in short chains or frameworks. The more felsic the magma, the more abundant and longer the chains.

A volatile-free silicate magma therefore consists of a mixture of negatively charged SiO_4 tetrahedra (isolated or in chains or frameworks whose complexity is greater for more felsic compositions) and positively charged metal ions that tend to interrupt the chains. The lack of more than purely local order in the structure of such a mixture allows it to behave as a liquid.

An important property of silicate magmas is that they tend to be slightly less dense than their solid equivalent. This means that magmas have a tendency to rise buoyantly. How freely magma is able to move, both at depth and when erupted at the surface, depends on its **viscosity**. This is a measure of resistance to flow, and is measured in units of $kg\,m^{-1}\,s^{-1}$*. Long chains hinder the free flow of a magma, so felsic magmas are more viscous than mafic magmas. The viscosities of mafic, intermediate and felsic magmas are typically of the order of 10^2, 10^3 and $10^5\,kg\,m^{-1}\,s^{-1}$ respectively (for comparison, the viscosities of water and porridge are about 10^{-3} and $10^2\,kg\,m^{-1}\,s^{-1}$). For a given magma, viscosity decreases slightly with temperature, but this is insignificant compared to the compositionally controlled differences in viscosity between magma types.

When volatiles are present in a magma, they occur as isolated molecules dispersed among the silica chains and metal ions. The higher the confining pressure, the greater the amount of volatiles that the magma can hold in this way. However, if the pressure falls, some of the volatiles will begin to exsolve.

* These units can also be expressed as 'pascal seconds' (Pa s), because $1\,Pa = 1\,kg\,m^{-1}\,s^{-2}$, so $1\,Pa\,s = 1\,kg\,m^{-1}\,s^{-1}$.

- ❑ Can you think of a natural situation in which the pressure experienced by a body of magma would fall?
- ■ The most obvious case is when magma is rising upwards.

The confining pressure will fall as the magma rises to shallower depths. This tendency for volatiles to exsolve at low pressures is the main driving force behind explosive volcanic eruptions.

Volatiles also play a role in influencing viscosity. Like positively charged metallic ions, volatile molecules can get between the chains of SiO_4 tetrahedra and hinder their polymerizing to form longer chains, thereby reducing the viscosity of the magma. The tendency of volatile molecules to disrupt chains also explains the lowering of melting temperatures when volatiles are present (Figures 4.2 and 4.3).

4.3 Migration of magma

Now let's consider how magma can collect into a body and rise upwards through the crust. Irrespective of whether melting is initiated by a rise in temperature, a change in pressure, or the addition of volatiles, partial melting begins by yielding a small amount of melt by the preferential melting of the starting material's most silica-rich minerals. The first melt will form at the boundaries between melting grains and their neighbours. Studies have shown that no matter how tiny the percentage of melt in a partially melting rock, the melt usually has the property of 'wetting' the grain boundaries, so that the melt is interconnected rather than occurring as isolated droplets (Figure 4.4).

You have seen that silicate melts are less dense than silicate rocks of the same composition. It is also the case that silica-rich rocks are less dense than silica-poor rocks. For both these reasons, melt that has begun to form by partial melting is buoyant and so has a tendency to rise upwards. However, melt is inhibited from moving by resistive forces. These comprise its own viscosity (the more viscous the melt, the harder it is to make it flow) and the resistance offered by the surrounding rock.

Studies have shown that unless melt is squeezed out by exceptionally strong tectonic deformation, resistive forces will prevent an initial 'grain boundary melt' escaping from its source until sufficient melting has occurred to produce a network of melt-filled veins occupying about 5% of the source rock. In other words, about 5% of the source rock must have melted. It can take up to about 5×10^4 years for this threshold to be reached, but then the melt can flow freely. If there is a dyke-like fracture available (perhaps because the crust is under tension), then magma can escape up it in a matter of a thousand years or less. On the other hand, if there is no easy pathway, the melt must collect into a body many kilometres across before it is capable of forcing its way upwards by means of its own buoyancy. This would require about 30% of the source to melt, and take in excess of 10^5 years.

Igneous rocks that were intruded more than about a kilometre deep in the crust commonly occur as bodies up to about 30 km across referred to as *plutons* (Block 1, Section 2.2). Plutons cannot be studied by direct means when they are forming, so geologists interpret what went on by studying ancient plutons exposed at the surface by erosion. The most important question to decide is how the pluton came to be in the place where it is found. Very often the surrounding rock, which is conveniently referred to as the *country rock* (Block 1 Section 2.2), is chemically unrelated to the intrusion and shows no sign of having had any melt extracted from it. Thus, we are left seeking an explanation of how a substantial body of magma could find its way into an alien position and make room for itself.

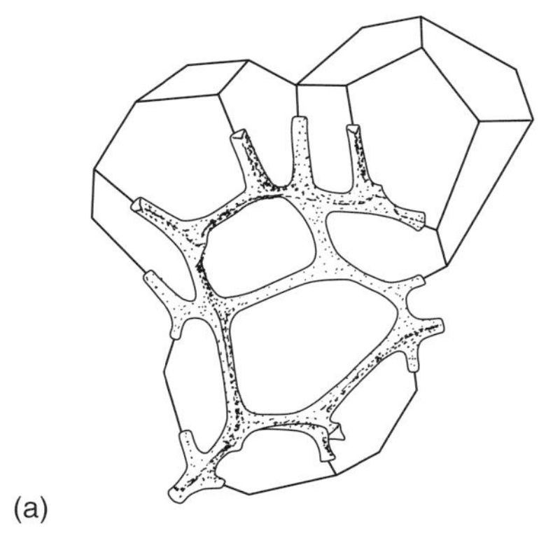

(a)

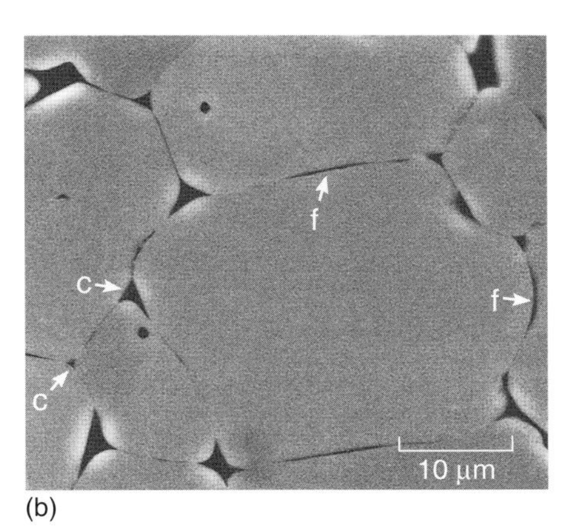

(b)

Figure 4.4 (a) Sketch showing a three-dimensional view of adjacent crystals in a rock in which partial melting has begun. Melt forms in pockets where three crystals meet, but pockets are connected by channels of melt running along grain boundaries. (b) Magnified image of a rock beginning to melt showing melt (dark) and crystals (light). Features labelled 'c' are melt channels along boundaries where three grains meet; features labelled 'f' are much thinner films coating grain surfaces.

You will see some evidence bearing on emplacement mechanisms for plutons in Section 5.1. In the meantime, draw on principles from the whole of Section 4 so far to answer Question 4.4.

Question 4.4 Suggest why partial melting of a mafic source rock is unlikely to yield a large body of felsic magma.

Large bodies of felsic magma can, however, result from partial melting of intermediate rocks, from which a significant volume of felsic melt can be generated before the average melt composition becomes intermediate.

Before looking at plutons in more detail, there are some important processes affecting the magma that you need to explore. As magma rises to shallower levels within the crust, it will lose heat to the surrounding rocks. Whether or not this cooling causes it to solidify depends on its starting temperature, the rate at which it loses heat as it rises, and the relationship of these to the phase boundaries on the relevant partial melting phase diagram. A magma body in the crust will usually keep rising until it has become almost entirely solid. Even a 'crystal mush' consisting of over 90% crystals and less than 10% melt can continue to rise. If a magma body solidifies completely without reaching the surface, it will form an intrusive igneous rock, whereas if it is erupted at the surface it produces a volcanic rock.

Various things can happen to a magma as it rises that will affect its composition. Its path in P–T space may cause it to cross the liquidus from the liquid only to the solid+liquid field. In other words, it will begin to crystallize according to the sequence of crystallization you met in Plate 6.5 of Block 2. If these crystals are carried along with the remaining melt until it has completely solidified, the composition of the eventual rock will be identical to that of the initial magma body, and the first-formed crystals will be phenocrysts.

❑ However, how will compositions be affected if the first-formed crystals become separated from the melt?

■ The escaping melt will be more felsic than the original melt, whereas the separated crystals will form a rock less felsic (more mafic) than the original melt.

Such physical separation of the crystal fraction from the melt fraction in a partly crystallized magma is a very important way in which the composition of a magma can change. The process is termed **fractional crystallization**. It can occur by the magma being squeezed out of the space between the crystals, or by settling of crystals if magma becomes stored for a long period in a magma chamber. The former is more likely if magma is migrating in small quantities, and the latter if it is migrating upwards as a pluton-sized body.

Another way in which magma composition can change is if the magma has enough heat to melt some of the rock it passes through. If this new melt is able to mix into the main magma body, then this will develop a new composition that is a weighted average of its initial composition and that of the new material that has been mixed in. This process is described as **assimilation**.

❑ Do you think it likely that a rising body of felsic magma would assimilate mafic rocks through which it was rising? (Refer to Figures 4.2 and 4.3.)

■ Mafic rocks begin to melt at a much higher temperature than felsic rocks, so a felsic magma is unlikely to be hot enough to melt a mafic rock.

The converse is more common, in that a rising body of mafic magma can assimilate rocks of intermediate and felsic composition, and a rising body of intermediate magma can assimilate rocks of felsic composition. In the process, the magma becomes more felsic than its initial composition. Both fractional crystallization and assimilation therefore cause the silica content of a magma to increase.

What creates the space to allow a large magma body to move upwards through the crust is something we shall examine in the next Section. However, we will note here that the heat available even in a large magma pluton is insufficient to assimilate more than a limited volume of country rock before the pluton itself would become solidified. Thus, assimilation cannot be the primary explanation for how plutons find room for themselves.

4.4 Summary of Section 4

- Any silicate rock consisting of more than one mineral melts over a range of temperatures and pressures. In a *P–T* phase diagram, the solidus and liquidus separate stability fields of solid only, solid (crystals) + liquid, and liquid only.
- Under hydrous conditions, or in the presence of any other volatile, both solidus and liquidus plot at lower temperatures for a given pressure (except for zero pressure).
- Partial melting of silicate rocks yields a melt that is more felsic than the starting material. This can become separated from the remaining crystals, leaving these behind as a residue that is less felsic than the starting material.
- Viscosity of magmas increases with silica content, because flow is hindered by chains of SiO_4 tetrahedra.
- Magmas tend to rise because they are less dense than solid rock. However resistive forces prevent migration until about 5% of the source rock has melted. Even so, a fissure is required to allow the magma to penetrate upwards except when a magma body large enough to punch its way upwards has been generated (requiring about 30% melting of the source).
- Volatiles have a greater tendency to exsolve (form bubbles) in magma as the pressure drops.
- A magma can become more felsic by fractional crystallization (if the crystals are removed) or by assimilation of rock it encounters during uprise.

4.5 Objectives for Section 4

Now you have completed this Section, you should be able to:

4.1 Explain how melting and crystallization occur with reference to anhydrous and water-saturated *P–T* phase diagrams.

4.2 Explain how melt migrates, tracing all the steps from grain boundary wetting through to uprise of magma bodies.

4.3 Describe how the viscosity of a melt depends on its composition.

4.4 Explain how the composition of igneous rocks depends on a combination of melting, crystallization and assimilation of country rocks.

In the next Section, we will look at various kinds of igneous intrusion. In the meantime, try the following question to test your understanding of Section 4.

Question 4.5 In continental areas, temperature usually rises with depth at a rate of 20–40 °C km^{-1}. Suppose the rate in a particular region is 30 °C km^{-1}.

(a) What would you expect the temperature to be at a depth of 30 km? (Assume that the surface temperature is 10 °C.)

(b) Plot this point on Figure 4.3, using the depth scale on the right of the diagram, and label this point A. If conditions are anhydrous, explain whether the physical state of felsic material will be solid, liquid or both.

(c) Suppose now that sufficient water is introduced to saturate the system. Explain what the physical state of felsic material will be now.

(d) Suppose the resulting magma rises towards the surface, and that it loses heat at an average rate of 5 °C km^{-1}. Plot a line starting from point A representing the path through P–T space followed by magma cooling at this rate as it rises, and with reference to this line describe what will happen. Speculate on whether or not any magma will reach the surface.

5 Intrusive rocks

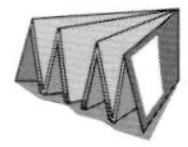

Study comment: *You will need your Ten Mile Map (N) and Lake District Sheet handy for Activity 5.1, the text following it, and Activities 5.2 and 5.3. Activity 5.4 summarizes part of this Section by means of a video sequence.*

In this Section, we will look at igneous bodies produced by intrusion of magma that has not reached the surface. We will begin with plutons, before turning to smaller intrusive bodies that are more characteristic of intrusion at shallow depths.

5.1 Plutons

Although most plutons are produced by solidification of large bodies of magma at depths of several km or even tens of km, it is quite common to find ancient plutons exposed at the surface.

❑ Can you suggest why?

■ They are exposed as a result of erosion.

Exposed plutons are especially common in areas where the crust has been thickened, because crustal thickening leads to isostatic uplift, and uplift encourages erosion of the surface. Pluton-rich levels of the crust are therefore commonly exposed in collision zones and at destructive plate boundaries, which are both settings where the crust is thickened. As you will see in Sections 7.4 and 7.5, these settings actually favour the generation of plutons in the first place.

So what do plutons look like? We will demonstrate this with examples of felsic plutons.

❑ Bearing in mind the depth at which it crystallizes, would you expect the grain size of a pluton to be fine, medium or coarse?

■ A pluton crystallizes deep down. Therefore it cools slowly and the crystals have time to grow to a large size. Plutons are thus coarse-grained.

❑ Bearing in mind its grain size, what rock name would you give to a pluton of felsic composition?

■ It should be called a granite, according to the classification scheme you met in Plate 6.12 of Block 2.

We will look at examples of granite plutons that were emplaced (i.e. ceased to rise independently) at different depths, and their relationships with the country rock. We will begin with 'deep' granite plutons emplaced at mid-crustal levels. At this depth (about 20 km), most continental crust has felsic-intermediate composition but the heat and pressure are generally sufficient for the rock to be regionally metamorphosed. The country rock around deep plutons therefore consists of felsic-intermediate gneisses and schists, similar to RS 4 and RS 26 in your Home Kit. You should remember from Section 8 of Block 2 that gneisses

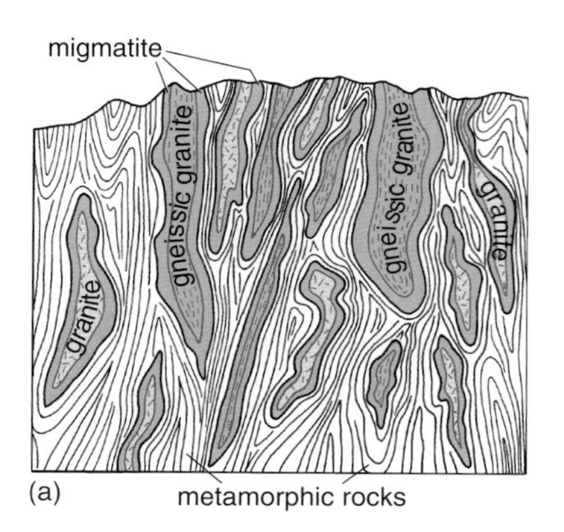

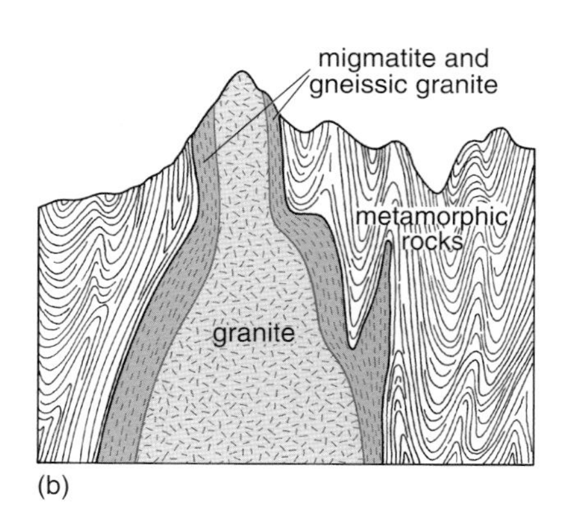

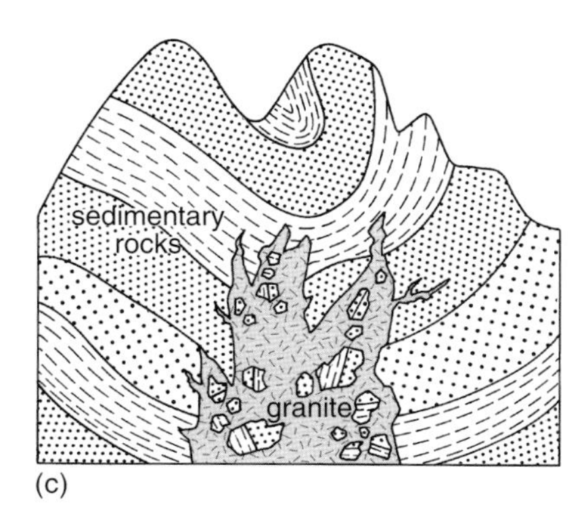

Figure 5.1 Schematic cross-sections showing generalized relationships between plutons emplaced at (a) deep (*c.* 20 km, i.e. mid-crust), (b) medium (*c.*10 km) and (c) shallow (*c.* 5 km) depths, in each case now exposed at the surface by erosion.

and schists have a well-developed metamorphic foliation, which you can see if you look at your specimens.

When a gneiss is subject to conditions close to those at which melting begins, ionic diffusion and a small amount of partial melting can turn its texture into that of a migmatite (Block 2 Plate 8.2). Pervasive fine vein-like networks of granite can be found in migmatite but not commonly within unrelated rock types found higher in the crust, and this is a strong indication that granitic magma usually collects into substantial amounts before rising much.

The deepest granite plutons tend to consist of elongate or tabular masses of granite aligned parallel to the metamorphic foliation of the country rock. Often the final stage of deformation of the country rock and the intrusion of the granite occur contemporaneously, so that the granite itself has a foliation imposed upon it. These relationships are portrayed in cross-section in Figure 5.1a.

If a granite reached a shallower level before ceasing to rise, it is more common to find a single large pluton rather than many small ones (Figure 5.1b). The country rock is still deep enough to be regionally metamorphosed, and the edges of the pluton still tend to be parallel to the foliation, rather than cross-cutting it. To use terms you met in Block 1, we can say that these plutons tend to be *concordant* to the foliation of metamorphosed country rock, rather than *discordant* with it. The explanation is that at this depth the country rock was hot enough (and therefore deformable enough) for a pluton-sized mass of granite to force its way upwards by pushing the country rock aside (possibly aided by assimilation). A pluton that has risen in this way is described as a **diapir**. A granitic diapir could rise from the lower crust to about 10 km below the surface in about 10^4–10^5 years, but at shallower depths the crust is not deformable on a sufficiently short time-scale to allow further diapiric rise. A pluton would be expected to solidify at this depth in about 10^5 years.

Plutons believed to have been emplaced at shallower levels in the crust (<10 km depth) can usually be seen to be *discordant*. Their edges cut across the fabric of the country rock, which at this depth is usually sedimentary bedding (Figure 5.1c). When intruding into cold, brittle crust, a shallow-level pluton cannot push the country rock aside and force its way up diapirically as may sometimes be the case deeper down. Nor can it simply 'melt its way upwards' by assimilating all the country rock in its path, because to do so would require more heat than the pluton contains. The most likely means of accumulating a large magma body at shallow depth is by magma rising up near-vertical fissures ('dyke ascent'). Continuous magma ascent through a 5 m wide fissure could supply an average-sized granitic pluton (5000 km^3) in about 10^4 years.

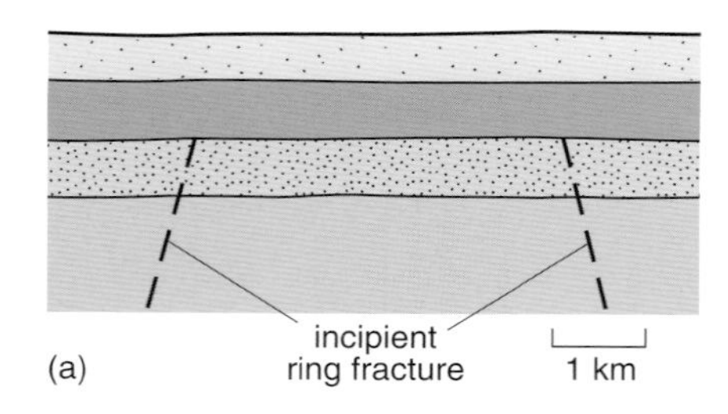

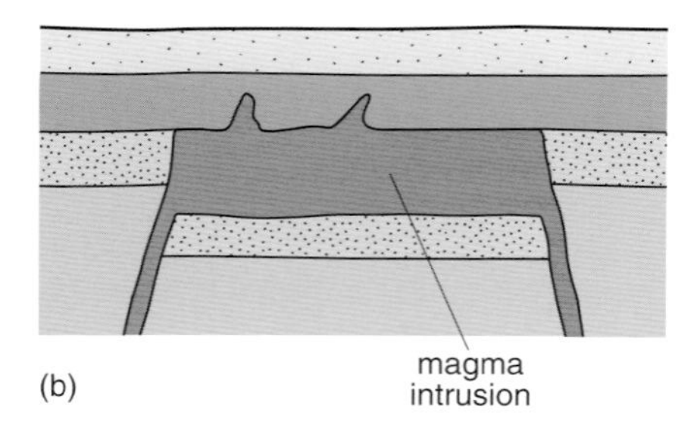

Figure 5.2 Cross-sections to show emplacement of a shallow-level pluton by subsidence of a cylinder of country rock bounded by a ring fracture. (a) A ring fracture forms, which would be approximately circular in plan view, but in this example does not reach the surface. (b) The cylinder of country rock bounded by the ring fracture subsides, and magma rises up to occupy an overlying magma chamber.

Given that assimilation cannot be a major factor, you should realize that space must somehow be generated within the crust for a pluton to occupy. As we have seen, the middle crust may be deformable enough to be pushed aside and deform around a diapir. However, plutons in the upper crust can often be seen to occupy spaces made available by fracturing and fault movement. One example is a **ring fracture** that allows a cylinder of country rock to subside, while the magma rises by dyke ascent up the fracture and collects in an overlying magma chamber where it solidifies as a pluton (Figure 5.2). Sometimes, the ring fracture may extend above the pluton, where it is manifested by features you will meet in Sections 5.2 and 6.4. In addition, particularly shallow plutons may be accommodated by a certain amount of updoming of the surface.

Field evidence suggests an additional process that can help to overcome this 'space problem'. Chunks of overlying country rock can be broken away and sink through the granite. These chunks have been drawn far too large in Figure 5.1c, but allowing for the exaggerated scale this is a fair enough representation of what can sometimes actually be seen in the field (Figure 5.3). A chunk of country rock enclosed within an intrusion is called a **xenolith** (*xeno*, pronounced 'zeno', is Greek for 'foreign', and the xenoliths are 'foreign rocks' in the sense that they do not belong in the intrusion).

Xenoliths are particularly common near the roofs and walls of intrusions, where they tend to be angular in shape and it is clear that they have been plucked away from the country rock by the magma. This process is described as **stoping**. Some xenoliths may sink to the bottom of the pluton, but the bases of plutons are rarely exposed so evidence is scarce. Xenoliths can sometimes be identified near the centres of plutons, where they tend to have more rounded shapes, and sometimes a recrystallized fabric, indicating that the heat from the surrounding magma was sufficient to cause them to become soft and mushy. In extreme cases, xenoliths are no more than ghostly dark patches, showing that they have become almost entirely assimilated into the granitic magma. It can be difficult to distinguish from these patches of other magma that have been incompletely mixed with the main granite magma.

Figure 5.3 Photograph of the roof zone of a pale granite pluton that has been intruded at shallow level into darker country rock.

When studying a pluton in the field, a geologist is seeing the culmination of long, complex and possibly multiple processes of melting, intrusion and crystallization. Many of the crystals, particularly the larger ones (phenocrysts), may have begun to grow at great depth and been carried upwards with the magma. It is a matter of some debate how fully crystalline a granite has to be before its upward rise is halted. Some estimates suggest that a crystalline mush with only 10% remaining melt would still be sufficiently mobile for emplacement mechanisms to operate. The final stages of crystallization are controlled by the fact that magma can hold a greater percentage of volatiles than can solid rock, so as crystallization proceeds the remaining melt becomes progressively richer in volatiles. The last fraction of a percent to crystallize may be exceptionally rich in volatiles (and in those metallic elements, such as lithium, that do not easily fit into common minerals) and form veins of coarse-grained pegmatite, which you met in Section 6.5 of Block 2. Alternatively, as you also saw in Block 2, sudden escape of the volatiles can trigger rapid crystallization of a fine- to medium-grained sugary-looking mixture of quartz and alkali feldspar referred to as aplite.

❑ You have seen that the heat from a granite can have extreme effects on xenoliths that find their way into the heart of the intrusion. But what effect, if any, would you expect to see in the country rock in contact with the granite?

■ You would expect heat to be conducted from the granite into the country rock, and this would cause contact metamorphism (Block 2 Section 8.1).

As you saw in Block 2, the zone of non-foliated, contact metamorphosed rock around a shallow granite pluton is described as its metamorphic aureole. This is alternatively known as a 'contact aureole' or 'thermal aureole' and can be of the order of a kilometre wide. You will study an example of contact metamorphism around a granite in Activity 8.2.

The heat of the pluton may also cause water within the crust to convect by **hydrothermal circulation**, in which hot water is expelled through the roof of the intrusion and colder water is drawn in through its sides. This fluid dissolves and reprecipitates certain elements, some within the pluton and some within the country rock, and the process may continue long after the pluton has solidified. Such movement of elements by hot solutions can cause further mineralogical changes in both the pluton and the country rock, an effect that is described as **metasomatism**. It may also lead to deposition of ore minerals, particularly in veins within and around the pluton (Block 2 Plate 6.11), sometimes in economic quantities.

Activity 5.1 (later in this Section) will require you to compare the characteristics of some granitic intrusions on the Ten Mile Map (N), and to see what you can infer about their depth and mode of emplacement. The Key to this map includes names of some igneous rock types that you have not yet met, so before progressing we must put your command of igneous rock names on a more complete basis. In Block 2 Section 6.7, you were introduced to a compositional classification of igneous rocks according to silica content as reflected by the abundances of quartz, feldspars, and mafic minerals. This is expressed in the scheme shown in Block 2 Plate 6.12 as a linear progression between felsic and ultramafic end-members. Although silica content is the *most important* variable in igneous rock composition, it is not the *only* one. Perhaps the next most important factor (and one which can vary independently of silica content) is the relative abundances of the alkali metals sodium (Na) and potassium (K).

❑ What common mineral group do you know that contains large amounts of Na and K?

- The feldspars; particularly the sodium-rich plagioclase feldspar albite ($NaAlSi_3O_8$), and potassium feldspar ($KAlSi_3O_8$).

Plagioclase feldspar, you may remember (Block 2 Section 4.6.2), exists as a solid-solution series of any composition between that of two extreme end-members: albite ($NaAlSi_3O_8$) and anorthite ($CaAl_2Si_2O_8$). A convenient basis for taking abundances of alkali metals into consideration for classification, yet which can be applied by visual inspection of hand specimens (for coarse-grained rocks, at least) is shown in Table 5.1. This requires the geologist to be able to distinguish and estimate the relative proportions of potassium feldspar, plagioclase feldspar and quartz.

Table 5.1 Classification of igneous rocks according to the percentage of quartz in the rock as a whole and the relative proportions of potassium feldspar and plagioclase feldspar (note that the columns and rows are not drawn to scale). Names for coarse-grained rocks are in upper case, with their fine-grained equivalents in lower case. Medium-grained rocks take the name of their coarse-grained equivalents with the term micro- added (e.g. microsyenite), except that a 'microgabbro' is usually called a dolerite. Gabbro is distinguished from diorite by the composition of its plagioclase feldspar, which is >50% albite in diorite and <50% albite in gabbro. (*Note*: all these rock types have other minerals present, notably mafic minerals, but this classification is based only on the quartz and feldspar content.)

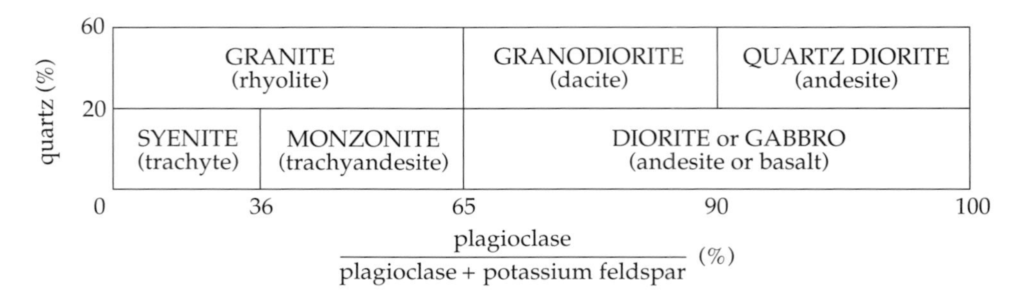

quartz (%)	plagioclase/(plagioclase + potassium feldspar) (%): 0–36	36–65	65–90	90–100
20–60	GRANITE (rhyolite)		GRANODIORITE (dacite)	QUARTZ DIORITE (andesite)
0–20	SYENITE (trachyte)	MONZONITE (trachyandesite)	DIORITE or GABBRO (andesite or basalt)	

Names from this Table that you have not yet met in this Course are syenite and its fine-grained equivalent trachyte (very rich in alkali metals but poor in quartz), monzonite and its fine-grained equivalent trachyandesite (poorer in alkali metals than syenite but not so poor as a diorite), and granodiorite and its fine-grained equivalent dacite (as rich in quartz as granite but poorer in alkali metals).

To reassure yourself that you can use Table 5.1 to classify igneous rocks, try Question 5.1

Question 5.1 What name would you give to igneous rocks containing the amount of quartz and relative proportions of feldspar given below, bearing in mind the grain size?

(a) Coarse-grained: 10% quartz, feldspar: 60% potassium feldspar, 40% plagioclase.

(b) Medium-grained: 30% quartz, feldspar: 30% potassium feldspar, 70% plagioclase.

(c) Fine-grained: 10% quartz, feldspar: 15% potassium feldspar, 85% plagioclase (of composition 30% anorthite, 70% albite).

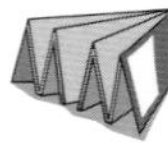

Activity 5.1

Now get out your Ten Mile Map (N) and do Activity 5.1. A more detailed map of one of the plutons in this Activity is shown as Figure 5.4.

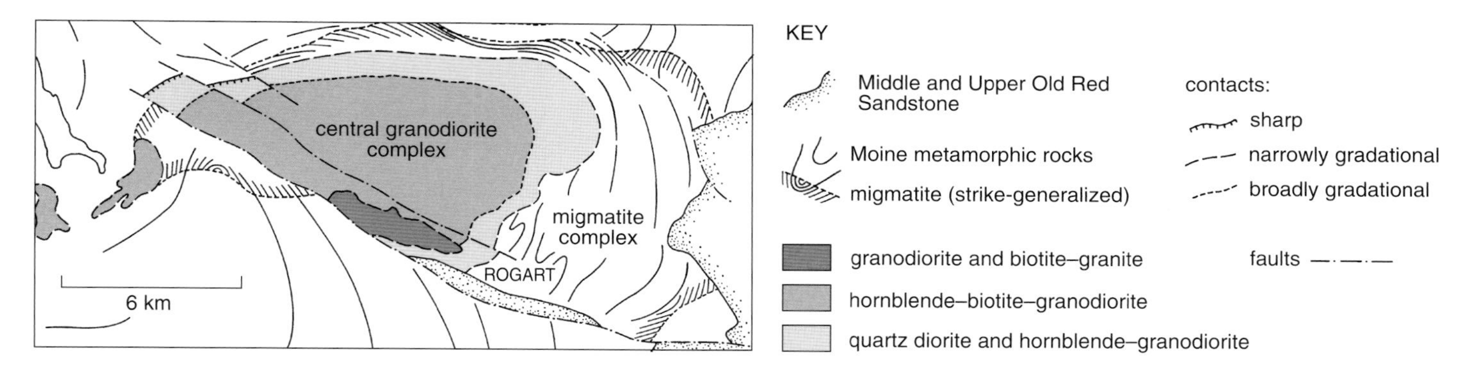

Figure 5.4 The Rogart intrusion, Sutherland, Scotland. For use with Activity 5.1.

The Rogart intrusion that you met in Activity 5.1 displays a feature characteristic of many granitic plutons, in that it consists of successive units intruded into one another (probably at intervals of the order of 10^4–10^5 years). As can be deduced from Figure 5.4, the first unit to be emplaced in the Rogart intrusion was the one mapped as quartz diorite and hornblende (a type of amphibole) granodiorite. The hornblende–biotite–granodiorite was then intruded into the centre of this mass, with a steeply dipping contact. This in turn was intruded by the unit mapped as granodiorite and biotite granite. It is common for the compositions of successive intrusions to evolve in this manner, from generally dioritic (intermediate) at first to generally granitic (felsic) towards the end. This probably indicates a greater amount of partial melting in the source region at first, with decreasing amounts of partial melting later on. Alternatively, it could also reflect a slower rate of movement for more felsic, and hence more viscous, melts.

Most geologists would regard the Rogart intrusion as a single pluton, even though it is a concentric complex constructed by the arrival of successive pulses of magma. Sometimes two or more neighbouring plutons overlap (the younger intruding the older) in a more complicated manner than the concentric pattern exemplified by the Rogart intrusion. Such a combination of plutons is referred to as a **batholith** (others may define it as any pluton or association of plutons exceeding 100 km^2 in surface area). A notable British example occurs beneath the Lake District where the Shap Granite that you met in Activity 5.1 is most simply interpreted as an upward protrusion from a batholith of early Devonian age that underlies the whole region, and which is exposed elsewhere as the Eskdale Granite and the Skiddaw Granite. You can identify these on your Ten Mile Map (N) and Lake District Sheet near NY (35) 1500 and NY (35) 3029 respectively. There is some doubt about their respective ages, and these three outcrops might be unrelated. The large granite body 10 km to the north of the Eskdale Granite is definitely older and not part of this batholith.

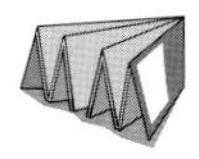

The floors of plutons and batholiths occur deep within the crust, and tend not to be exposed. On most cross-sections (such as those shown below your Lake District Sheet), you will see a pluton or batholith displayed as a steep-sided body whose base is conveniently ignored because it lies below the greatest depth shown on the cross-section. This reflects the fact that what happens at the base is genuinely not known in most cases. However, evidence from seismic studies (performed after your Lake District Sheet was compiled) suggests that the Lake District batholith is actually a series of kilometre-thick sheet-like intrusions with upward projections constituting the individual plutons.

Our account of plutons has concentrated on granites and related rock types. Mafic magmas tend not to form such big intrusions. With some notable exceptions, the largest tend to be about 10 km across, and can often be proven to be sheet-like or downward-tapering in form. In the next Section, we will look at intrusive bodies small enough for their complete shape to be determined by field mapping. You will need to keep your Ten Mile Map (N) and your Lake District Sheet handy.

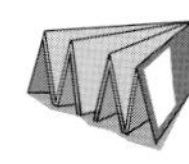

5.2 Minor intrusions

A minor intrusion is any intrusion too small to be classed as a pluton. Deeply eroded regions of crust expose evidence that small intrusions can be emplaced at great depth. However, the minor intrusions that concern us here are those emplaced at shallow depth, within a few kilometres of the surface, because these very often reveal links between plutonic and volcanic activity.

You met two kinds of minor intrusion in Block 1: dykes and sills.

❑ What is the difference?

■ A dyke is a discordant curtain-like intrusion, intruded in a near-vertical plane, whereas a sill is a generally horizontal sheet-like intrusion that is mostly concordant with the bedding of the strata it intrudes.

The kilometre-thick sheets proposed above for the Lake District batholith could be regarded as sills on a very large scale. However, sills are generally both smaller and shallower.

❑ Bearing in mind the small dimensions and shallow emplacement depth of minor intrusions, would you expect their grain size to be fine, medium or coarse?

■ Magma emplaced at shallow depth is surrounded by relatively cool rocks of the uppermost crust. When the magma body is small, as in a minor intrusion, it will lose its heat relatively quickly, and so minor intrusions generally have fine or medium grain size.

This means that the rock types of many minor intrusions should be named according to the medium grain size names in Block 2 Plate 6.12. For example, the rock type of a dyke of mafic composition and predominantly medium grain size should be called dolerite. However, the body itself may be referred to as a 'mafic dyke' or 'basaltic dyke'.

Sills and more especially dykes are associated with volcanoes, and there are other kinds of minor intrusions related to the magma 'plumbing system' within and below a volcano. Some information can be deduced about these by studying active volcanoes, but minor intrusions are best known from areas where erosion has exposed the roots of extinct volcanoes. Typical minor intrusions below a volcano are illustrated in Figure 5.5, and described on the next page.

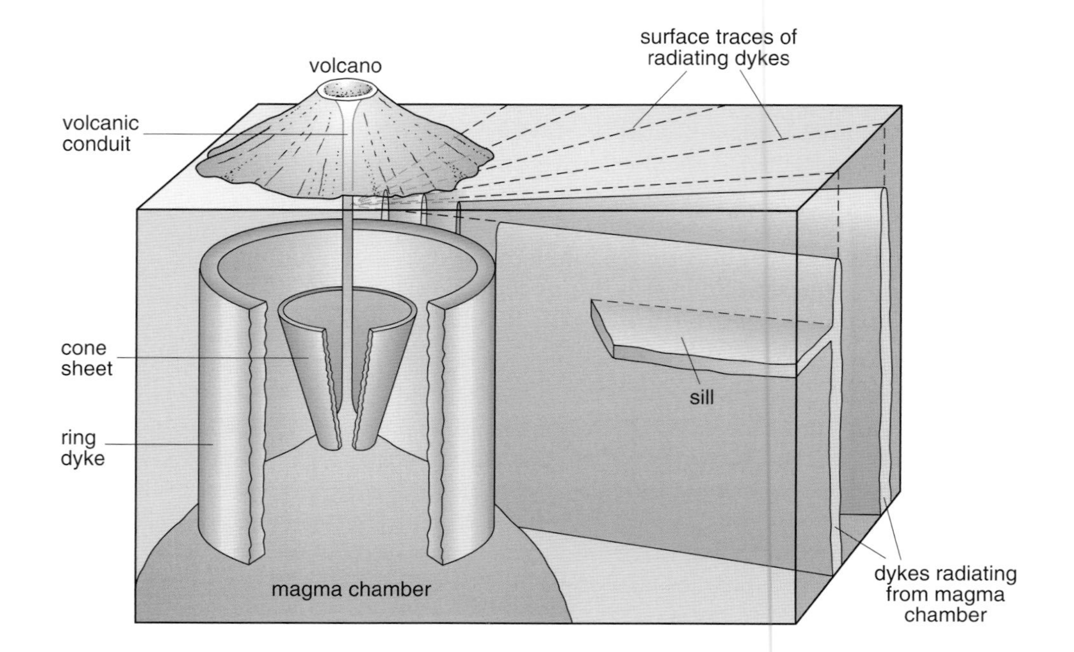

Figure 5.5 Diagrammatic three-dimensional composite view of minor intrusions commonly occurring beneath a volcano. (See text for discussion.)

The pipe feeding a volcano is described as a **volcanic conduit**. When the igneous rocks filling an extinct conduit are exposed by erosion of the surrounding material, they are described as a **volcanic plug** (or sometimes 'volcanic neck'). The conduit passes upwards from a plutonic intrusion representing the magma chamber. This chamber may supply several minor intrusions in addition to the conduit, and most of these are dykes. Commonly dykes occur as vertical planar bodies up to a few metres wide radiating outwards from the conduit and magma chamber; these are described as **radial dykes** or a radial dyke swarm. In the absence of external influences, radial dykes are symmetrically distributed (Figure 5.6a), but commonly dykes are deflected into a preferred orientation controlled by tension in the crust (Figure 5.6b). A set of parallel dykes that are not clearly radial from a common centre is referred to simply as a dyke swarm.

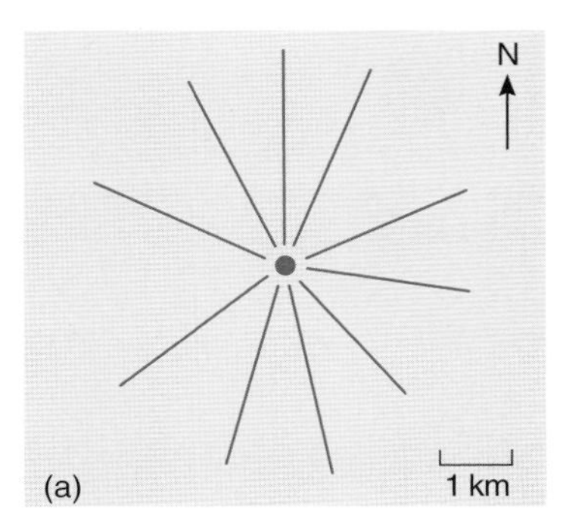

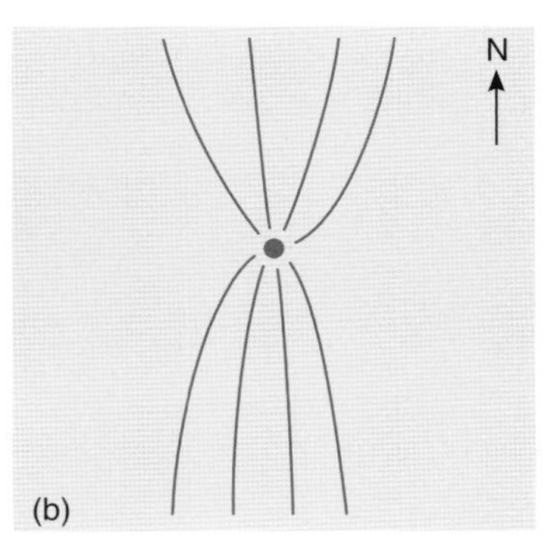

Figure 5.6 The outcrop pattern of radial dykes around a volcano in the absence of regional stress (a), and with east–west extension (b). Radial dykes may reach the surface of the flanks of an active volcano (in which case they may act as sources for subsidiary eruptions), but are more clearly seen after a period of erosion.

Activity 5.2

Now do Activity 5.2, in which you are invited to examine the outcrop patterns of some British radial dykes.

Dykes do not always reach the surface when they are emplaced, and if they do this may only be at one or two points along their length. For example, even after considerable erosion of the original overburden, the Cleveland Dyke outcrops discontinuously between a point near Carlisle (NY (35) 4050) and the Cleveland Hills but has no major outcrops between Carlisle and its source beneath Mull. A dyke that does not reach the surface may get narrower towards its top and have no surface expression, or there may be a partially open fracture above the dyke that is apparent at the surface. However, where erosion exposes an ancient dyke, it may be more resistant to erosion than the country rock and so appear as a positive topographic feature (Plate 5.1). The geometry of radial dyke swarms suggests that the dykes are fed from the central magma chamber, rather than from below along their entire length. Such dykes must therefore be emplaced by the lateral drawing of magma into a fissure as it propagates away from its source, which is estimated to occur at a speed of about 1–5 $\mathrm{m\,s^{-1}}$.

Cone sheets and ring dykes are dykes with approximately circular outcrop patterns (so they depart from the definition of a dyke as a planar body). **Cone sheets** converge downwards towards the top of the magma chamber. They are usually narrow features a few metres or less in width. Conversely, a **ring dyke** dips steeply outwards and is typically anything from a hundred metres to a couple of kilometres wide. The geometry of these two kinds of minor intrusion is illustrated in Figure 5.7; study this and then attempt Question 5.2.

Question 5.2 Bearing in mind that emplacement of dykes of any kind requires displacement of the country rock, how do the blocks of rock labelled A on Figure 5.7 move relative to the magma chamber during emplacement of (a) cone sheets, (b) a ring dyke? In each case, decide whether this implies that the minor intrusion was created by forceful injection or passive intrusion of magma.

Ring dykes may be related to deeper ring fractures of the kind you met in the previous Section in the context of subsidence of the floor of a magma chamber, but they are more clearly associated with surface subsidence and eruption of displaced magma, as you will see in Section 6.4.

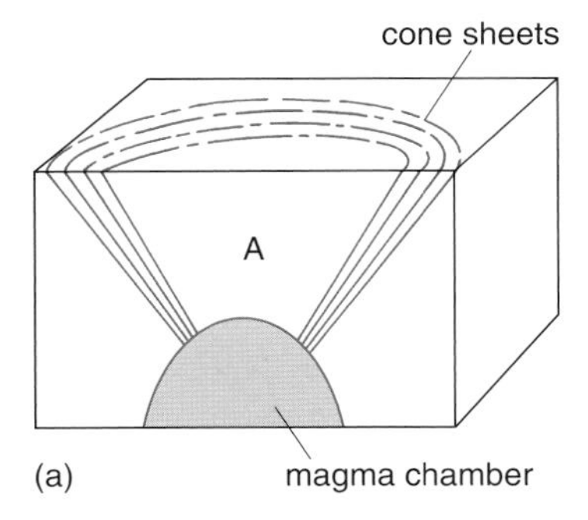

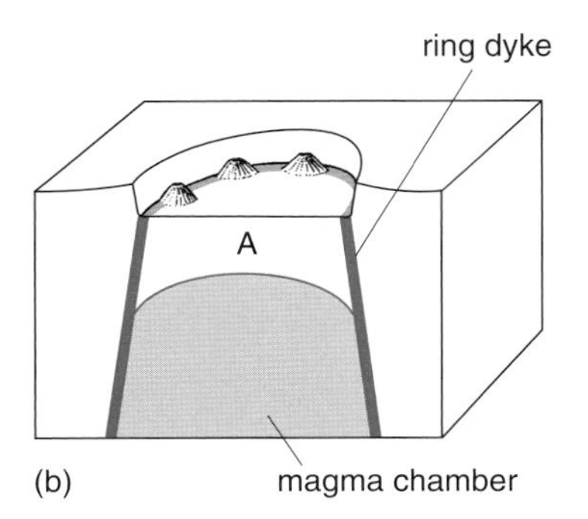

Figure 5.7 Diagrammatic cross-sections through a series of cone sheets (a) and a ring dyke (b). The circular depression bounded by the ring dyke in (b) is an example of a volcanic caldera, which will be defined in Section 6.4.

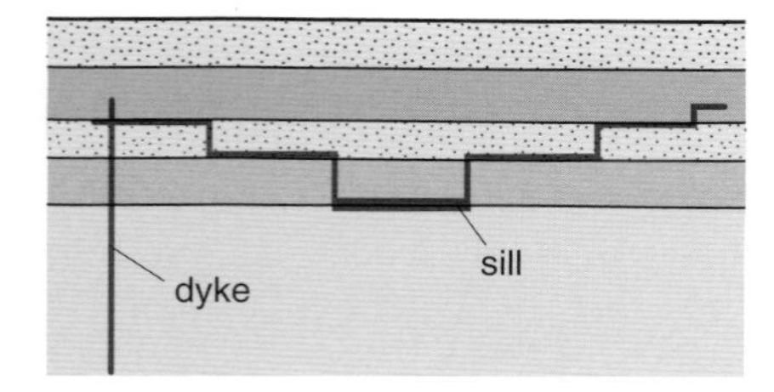

Figure 5.8 Cross-section showing the typical relationship between a sill (here fed by a dyke on the left) and horizontal strata intruded by it. The sill is generally concordant, but is locally discordant where it cuts up or down between bedding planes.

Sills may be up to hundreds of metres thick, and usually occur only at very shallow levels within the crust. They are usually fed by dykes, and are common where mafic magma is injected into horizontal or gently dipping sedimentary strata within a kilometre of the surface. Figure 5.8 shows a schematic cross-section through a sill showing its generally concordant but locally discordant nature.

A notable feature of sills is that the slight contraction a sill experiences as it continues to cool after solidification may cause a polygonal pattern of fractures called columnar joints to develop (Plate 5.2). The columns defined by the joints run perpendicular to the margins of the sill. Columnar joints are also sometimes found oriented perpendicularly to the margins of dykes.

Activity 5.3

You should now do Activity 5.3, which introduces you to some of the characteristics of a sill that can be inferred from a map.

We noted in Section 5.1 that the heat from a pluton is sufficient to develop a contact metamorphic aureole in the adjacent country rock, which is most apparent at shallow depths where the crust had been relatively cold until the arrival of the pluton. However, minor intrusions do not generally cause significant contact metamorphism of the rocks they invade. The temperature of the magma may be the same as in a pluton, but the total amount of heat escaping from a minor intrusion is not enough to heat the country rock to a sufficient temperature. Thus, sedimentary rock may be baked and become hard and brittle in a zone a metre or so wide against the contact with a dyke or sill, but growth of new, metamorphic, minerals is uncommon.

On the other hand, the loss of heat from a dyke or sill that has been emplaced into cold country rock is sufficiently rapid to express itself *within* the minor intrusion. The magma at the margin is cooled so rapidly that at most only fine-grained crystals have time to form, and the texture right at the contact is often glassy. Minor intrusions are therefore characterized by **chilled margins**, where the medium grain size of the interior of the body becomes progressively finer (or even glassy) over a centimetre- or metre-wide zone. A thin section through the chilled margin at the base of the Whin Sill is shown in Plate 5.3. The field of view is too small to show how grain size varies with distance from the margin of the sill, but you can see a similar effect in Figure 5.9, which shows thin sections of two samples from the Cleveland Dyke, collected from County Durham where the dyke is about 20 m wide. Look at these, and then attempt Question 5.3.

Question 5.3

(a) Decide whether the average grain sizes of (a) and (b) in Figure 5.9 are coarse, medium, fine or glassy.

(b) Which of the two thin sections comes from closer to the margin of the dyke?

(c) How would you explain the presence of the relatively long (0.3 mm) plagioclase feldspar crystal visible in each specimen?

In contrast to minor intrusions, shallow plutons rarely exhibit a chilled margin. This is because heat passing outwards from the interior of the pluton is usually sufficient to keep the rate of cooling slow even at the edge. In fact, if this were not so, it would be difficult to imagine how a thermal aureole could be developed in the country rock around a pluton.

(a)

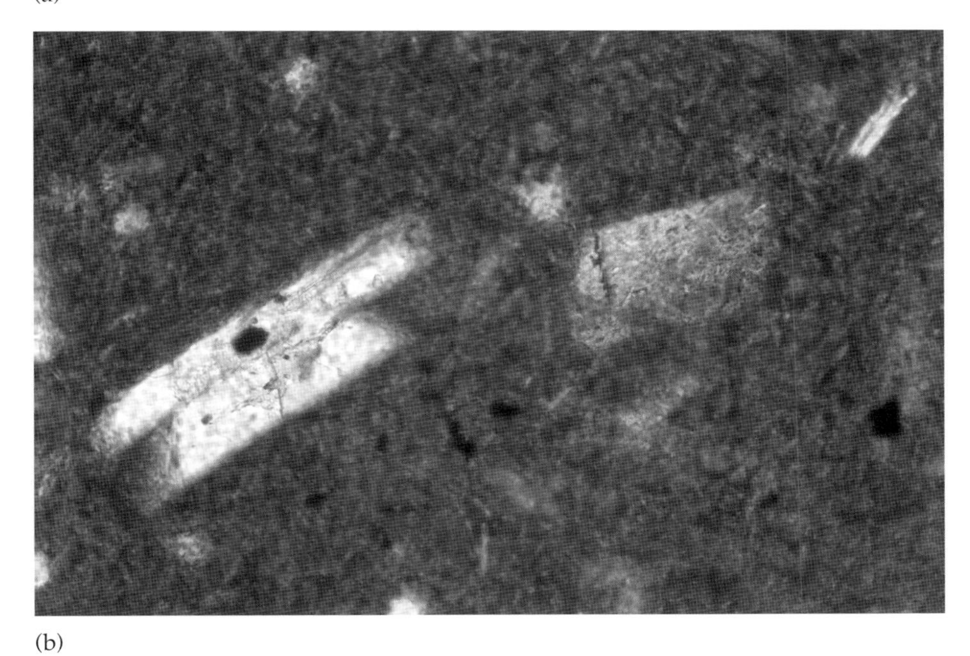

(b)

Figure 5.9 Thin sections from the Cleveland Dyke (plane-polarized light). The field of view is 0.7 mm from side to side. For use with Question 5.3.

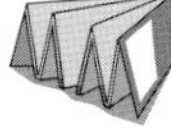

Activity 5.4

You should now do Activity 5.4, which is based on a video sequence on DVD 2 of some minor intrusions on the Ardnamurchan peninsula, Scotland.

5.3 Summary of section 5

- Plutons tend to be coarse-grained and lack chilled margins, whereas minor intrusions are fine- or medium-grained and have chilled margins.
- Minor intrusions have little or no metamorphic aureole, but one can usually be distinguished around a shallow pluton.
- Deeper plutons tend to be concordant with the structure of their country rock, whereas shallower plutons tend to be discordant.

- Diapiric rise of plutons is not usually effective at <10 km depth. Plutons emplaced at shallower levels are probably fed from below by dykes or along fault planes, in which case fault movements create the space occupied by the pluton, with some assistance from updoming and stoping.
- In the case of minor intrusions at even shallower depths, dykes fill extensional fractures, whereas the emplacement of sills is permitted when magma pressure can lift the overlying rocks and allow the sill to spread along a bedding plane.

5.4 Objectives for Section 5

Now you have completed this Section, you should be able to:

5.1 Recognize plutons and minor intrusions on geological maps and cross-sections, giving reasons for your diagnosis and suggesting, if appropriate, how they might be related.

5.2 Suggest reasoned explanations for the relationships between an intrusion and its country rock.

5.3 Describe probable emplacement mechanisms for different kinds of intrusion.

5.4 Classify igneous rocks on the basis of felsic mineral content.

Now try the following questions to test your understanding of Section 5.

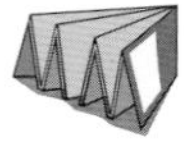

Question 5.4 On your Ten Mile Map (N), identify the following igneous intrusions, and then identify them as pluton or minor intrusion, concordant or discordant, and specify the age of each as fully as possible on the evidence of the map: (a) Unit 35 between NS (26) 8718 and NY (35) 3388, (b) Unit 34 centred at NX (25) 5570.

Question 5.5 The rock forming a fine-grained dyke has 10% quartz and four times as much potassium feldspar as plagioclase feldspar. What rock name should you give it, according to Table 5.1?

Question 5.6 Figure 5.10 shows two granites and an associated set of dykes in an intrusive complex of early Devonian age in western Scotland. You should be able to identify the area on your Ten Mile Map (N), but that map does not subdivide the granite. The rock type of the dykes is a porphyritic microdiorite (this is the meaning of 'porphyrite' in the key for the Ten Mile Map). Study Figure 5.10 carefully, and then:

(a) Decide, giving reasons, which of the following features you can see on the map: batholith, concentrically intruded pluton, ring fracture, ring dyke, cone sheet, radial dykes, dyke swarm.

(b) Describe, giving reasons, the order in which the two granites and the dykes were intruded.

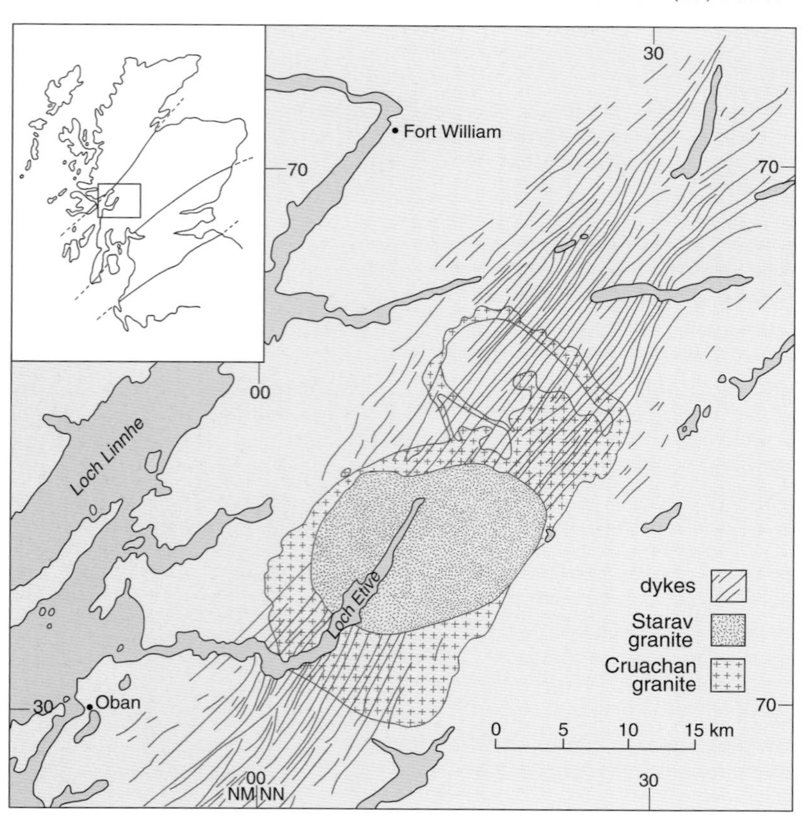

Figure 5.10 Map for use in Question 5.6.

6 VOLCANIC ROCKS

There are two videos associated with this Section, containing sequences chosen to illustrate the kinds of lava flow and pyroclastic eruption described in the text, a long Activity involving a DVD-based experiment on lava flows, and a rock specimen and thin section-based Activity to look at the water content of the igneous rocks in your Home Kit.

We will now look at what happens when igneous rocks are erupted, beginning with **effusive eruptions**, in which magma flows out of the ground. Magma flowing at the surface is described as lava, and the same term is used to describe the rock after it has solidified. A mass of lava emplaced by a single effusive episode is described as a **lava flow**, whereas a series of overlapping lava flows from a common source is described as a **lava flow field**. In this Section, we will deal in turn with lava flows on land and under water, then look at pyroclastic eruptions before exploring how combinations of eruption processes result in different kinds of volcanoes.

6.1 LAVA FLOWS ON LAND

Lava may be erupted either from a point described as a **vent**, for example where a volcanic conduit reaches the surface, or along a linear crack described as an **eruptive fissure** such as where a dyke reaches the surface. The form taken by a lava flow depends very much on how the lava actually moves. Factors controlling this include the steepness of the slope (if any) upon which the flow is emplaced, the rate at which lava is supplied at the source (the **effusion rate**), and the cooling rate. There is another factor that is more important than either of these.

❑ Can you think what this is?

■ The form of a lava flow must depend very strongly on how *freely* the lava is able to flow, in other words upon its viscosity.

Viscosity decreases with increasing volatile content and temperature, and increases with the proportion of crystals contained in the melt. However, as you saw in Section 4.2, bulk composition is a much more important factor. Lower silica contents give magma (and hence lava, which is what we call magma at the surface) very much lower viscosity. The eruption of ultramafic lavas is virtually unknown today, so we will confine our observations to mafic, intermediate and felsic flows. The lower viscosity of mafic lava means that it can spread more thinly, and flow down gentler slopes, than intermediate and felsic lava.

When young or active basalt lava flows are examined in the field, for example in Hawaii where they are widespread, a wide variety of surface morphologies can be seen. At one extreme, the surface takes the form of rough clinkery blocks, whereas at the other it consists of a less broken surface that may be wrinkled and draped to give the appearance of coiled rope (Plate 6.1). These two surface textures are known to volcanologists by the Hawaiian terms **a'a** (the pronunciation of which – *ah-ah* – is said to resemble the cries of pain uttered when walking barefoot over such a loose and jagged surface) and **pahoehoe** (pronounced *pa-hoey-hoey* or *pa-hoy-hoy*).

The factors controlling the development of lava surface morphology are complex. Observations of active lava flows and laboratory experiments using hot wax to simulate lava show that pahoehoe surfaces can form only when the rate of flow is slow enough to allow the surface to chill and produce a solid but pliable crust a few millimetres thick over most of the flow surface, whereas a'a texture is originated where flow is too rapid to allow a coherent crust to survive. A range of morphologies is recognized corresponding to flow speed as shown in Figure 6.1.

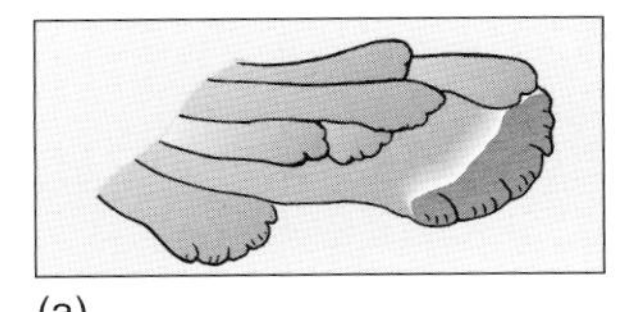
(a)

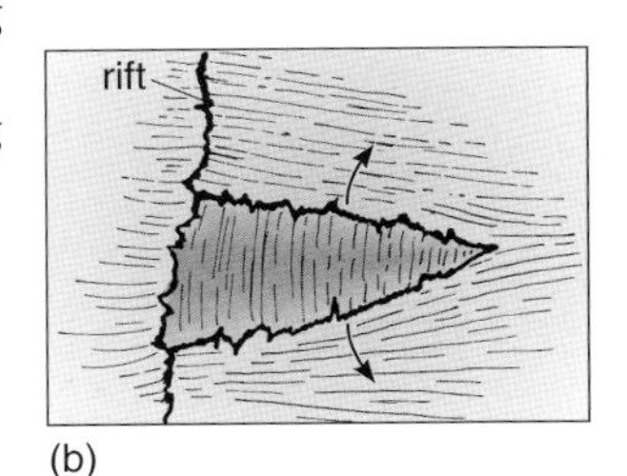

(b)

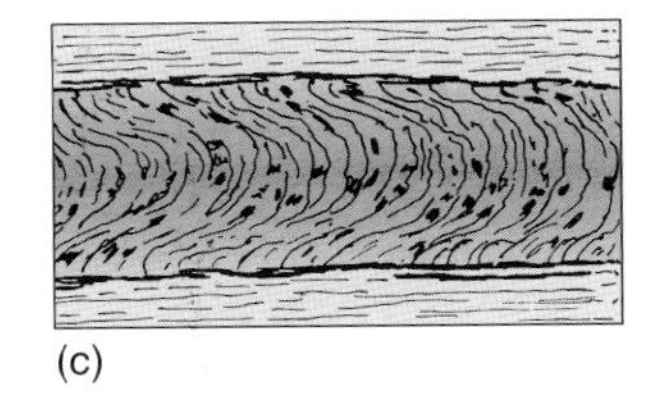
(c)

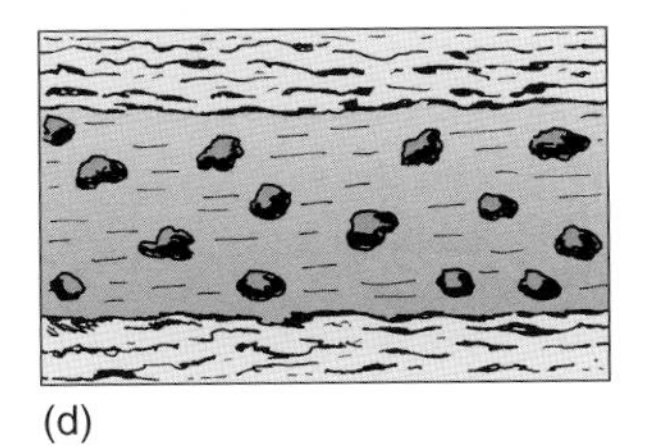
(d)

Figure 6.1 Series of basaltic lava flow surface morphologies, generated with increasing flow rate from (a) to (d): each sketch is several metres across. (a) Lobate or toey pahoehoe; (b) lineated sheet flow or toothpaste pahoehoe; (c) ropey pahoehoe; (d) jumbled sheets (which is one way that a'a can be formed). See text for discussion.

(a)

(b)

Figure 6.2 (a) Cross-section through a fresh sample of pahoehoe lava (Hawaii), showing vesicles. In this specimen, the size of the vesicles decreases upwards. (b) Large vesicles in the interior of an a'a flow (Etna).

In Figure 6.1a, molten lava is supplied very slowly to the flow front. The chilled surface crust is elastic, and inflates to form a metre-scale lobe or toe as lava enters it. The crust is cooling and thickening all the time, and eventually it can stretch no further so it ruptures, and a new toe develops from the underside of a previous one. This morphology is described as lobate pahoehoe, or more euphoniously as toey pahoehoe. In Figure 6.1b, the lava supply rate is faster so the crust has to stretch more rapidly, giving it a striated (streaky) texture, interrupted where the crust has been torn open. This morphology is described as lineated sheet flow pahoehoe, or sometimes toothpaste pahoehoe (Plate 6.2). When the interior of the flow is moving faster, drag on the underside of the crust pleats the surface over upon itself, resulting in the classic ropey pahoehoe morphology (Figure 6.1c, Plate 6.3 and Block 2 Plate 6.8a). Even faster flow stretches the surface apart into isolated chilled blocks within an open molten channel (Figure 6.1d); when the flow rate decreases further downstream, these blocks pile up into the jumbled mass known as a'a (Plate 6.4).

On close inspection, basaltic lava can often be seen to contain tiny bubbles, usually a few millimetres across, known as **vesicles** (Figure 6.2). These are created by the exsolution of volatiles when the confining pressure on the magma was removed (i.e. when it reached the surface) but which were unable to escape before the flow solidified. Vesicles may be particularly large and abundant in a'a, and contribute to the clinkery appearance of its surface.

Flows of intermediate and felsic composition are too viscous to develop pahoehoe textures; instead their surfaces break into slabs or blocks, like a larger scale version of a'a but with fractured, less clinkery, surfaces (Figure 6.3). These are called **blocky flows**. Generally speaking, the more viscous the flow, the larger the blocks.

Figure 6.3 The blocky surface of an andesite lava flow.

A'a flows and blocky flows move in a similar fashion, illustrated in Figure 6.4. Rubbly material from the flow top cascades down the flow front and is overridden like the treads of a caterpillar tractor. When a flow that has been emplaced in this manner is seen in cross-section, its interior is usually intact (though vesicular) whereas both its upper and lower surfaces are rubbly.

A notable feature of blocky flows, a'a flows and the more swiftly flowing varieties of pahoehoe is that the central part of the flow moves faster than its sides, and develops into a channel hemmed in by a low wall (described as a levée) of solidified lava or rubble on either side. If the channel becomes temporarily blocked or the effusion rate increases so that a levée is overtopped at some point, lava will spill out sideways, giving the lava flow as a whole a lobate outline (Plate 6.5).

The effusion rate usually declines during the course of an eruption, and this gives the surface of a channelized lava flow the opportunity to become solidified, while molten lava continues to move below its frozen roof. As the lava supply continues to dwindle, the level of lava within this **lava tube** falls. The tube may become wholly or partly drained (Figure 6.5), and portions of the roof may collapse to reveal the interior at points along its length. Many flow fields of pahoehoe are fed from lava tubes, with lava being squeezed up through cracks in the roof of a tube to form slow-moving pahoehoe on the surface.

When felsic lava is erupted, it cannot flow far because of its extremely high viscosity. A single flow of rhyolite or dacite is therefore dome-like and may exceed 100 m in thickness (Figure 6.6). Rhyolite often has a quenched glassy texture, and is then described as **obsidian**, which you met in Section 6.1 of Block 2.

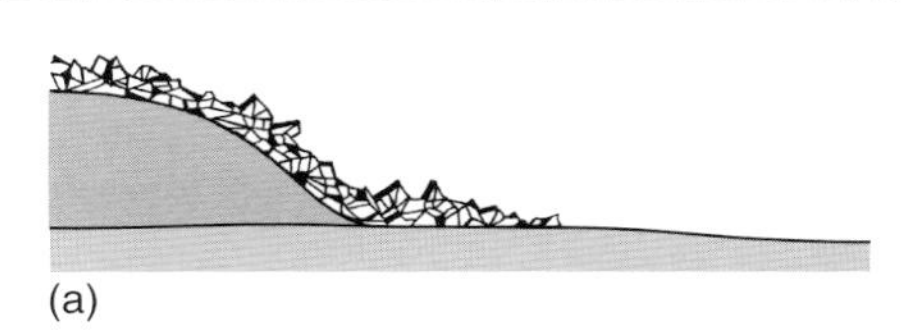

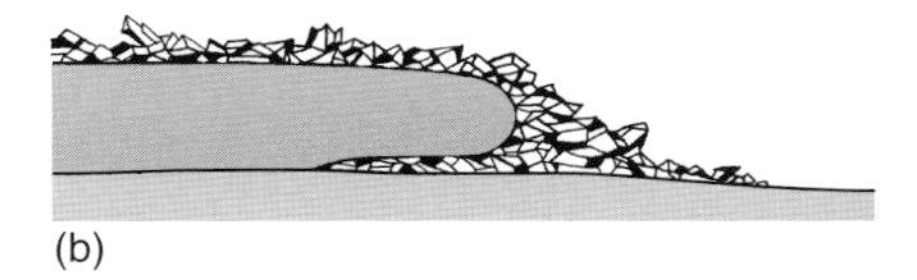

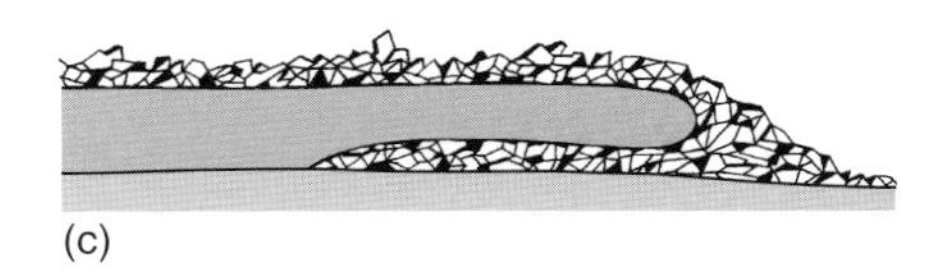

Figure 6.4 A blocky (andesite) or a'a (basalt) flow, seen in cross-section and advancing from left to right. Such flows are typically from one metre to several metres in thickness, but can be much thicker. The interior of the flow is molten, but the surface is rubbly. In (a), rubble falls from the front of the flow, and is progressively overridden (while fresh rubble continues to be shed at the front) in stages (b) and (c). Because its composition makes the molten interior of a blocky flow more viscous, it moves much more slowly than an a'a flow.

Figure 6.5 A lava tube, more than 10 km long, in Hawaii.

Figure 6.6 An image from space showing a 40 km wide region in the Andes. On the left is a 40 km^3 rhyolite flow, which flowed from north to south. Its front (in shadow) is 300 m high. The ridges on its surface are hundreds of metres apart. In detail, the surface consists of 10 m-sized blocks (too small to see on this image). To the right is a flat-topped dome of dacite that is presumed to have oozed out radially from a vent hidden below its centre.

Figure 6.7 Pillow lava, here seen partly in cross-section, which forms only when basalt is erupted under water. This sequence is the right way up, as can be inferred from the way in which the undersides of later pillows have moulded themselves to fill the cuspate cavities between the bulbous tops of underlying pillows. (Oman ophiolite, Arabia.)

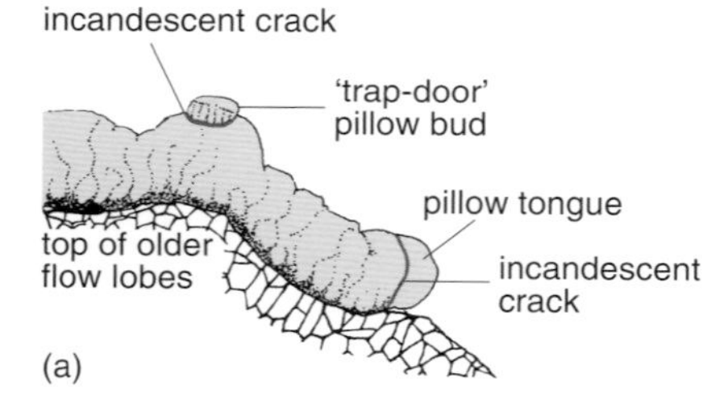

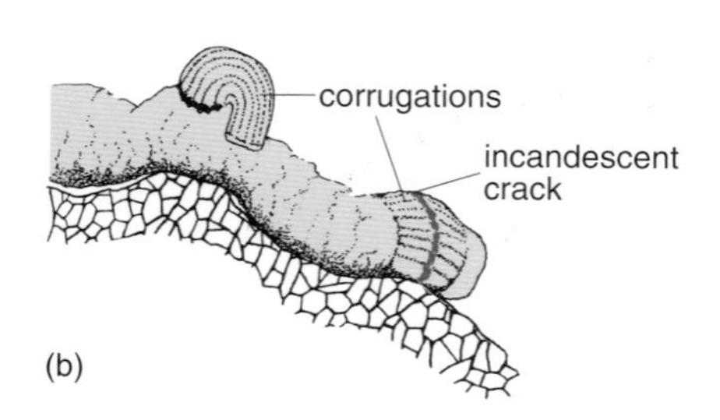

Figure 6.8 Stages in the development of pillow lava. (a) A pillow tongue growing down the slope is fed from a source to the left. Incandescent cracks in its skin open up as the pillow is inflated by the continued injection of lava. (b) The front of this pillow is pushed forward as an incandescent crack opens, while at the same time a new pillow begins to form where lava escapes from a 'trap-door' crack in the roof of the original pillow.

Whatever the morphology of a flow surface, or the nature of its base, if the flow is more than a few metres thick its interior is likely to cool as a coherent unit. As in the case of a sill, thermal contraction during cooling can often give rise to spectacular columnar joints as you saw in Video Bands 4 and 5 (see also Block 2 Figure 6.1 and Plate 6.1).

Question 6.1 What grain size(s) would you expect to encounter in a lava flow? Give your reasons. (*About 100–150 words*)

6.2 Underwater lava eruptions

When basalt is erupted under water, it can develop a flow morphology described as **pillow lava**, which is never produced on dry land (Figure 6.7). Pillow lavas form at slow effusion rates under water that on land would result in pahoehoe morphology.

❑ Can you think of a factor that could account for the special characteristics of basalt lava flows under water?

■ Slopes and effusion rates under water can be the same as on land, and viscosity must be identical because this question relates only to basalts. However, under water the rate at which the surface of a lava flow is cooled could be very different to the cooling rate on land.

In fact, water both *absorbs* and *transports* heat much more efficiently than air, so upon contact with water the skin of a lava flow is cooled extremely rapidly. Molten lava confined within such a skin produces a series of pillow-like or bolster-like lobes that are about 0.2–2 m in diameter. A pillow grows when the pressure of lava filling it ruptures the skin, which may lead either to the lengthening of the original pillow, or the budding off of a new one. Some of the processes that have been seen in progress are illustrated in Figure 6.8.

Although pillow lavas are characteristic of basalt eruptions under water, morphologies similar to pahoehoe and even a'a are formed if the effusion rate is high enough. It seems, therefore, that the spectrum of basalt flow morphologies on land is incomplete because the cooling rate in air is not sufficient to produce pillows, however slow the effusion rate.

Because the skin of a submarine lava flow is chilled so rapidly, it tends to flake away even as it forms, producing glassy fragments described as **hyaloclastite**. These may collect in the spaces between pillows, where they weather away rapidly to clay minerals. In more extreme cases, an entire flow may break into hyaloclastite fragments to give a hyaloclastite rock.

Intermediate and felsic lavas are too viscous for their morphology to be controlled by the glassy crust produced by eruption under water, so they tend not to form pillows. If a massive flow is formed, it usually looks rather similar to an equivalent flow on land, and the main effect of water is to encourage fragmentation which in extreme cases results in degradation of the flow to a pile of rubble. Although produced non-violently, hyaloclastite is fragmentary and so can be regarded as a pyroclastic rock (Block 2). We will look at more conventional pyroclastic rocks in the next Section, but first you are invited to strengthen your understanding of lava eruptions and how lava flows move by means of a couple of Activities.

Activity 6.1

First, do the DVD-based Activity 6.1, which illustrates much of what you have learned so far in this Section.

Activity 6.2

When you have finished Activity 6.1, you should perform the DVD-based Activity 6.2 in order to investigate some of the factors controlling the shapes of lava flow fields.

6.3 PYROCLASTIC ERUPTIONS

You saw earlier how volatiles coming out of solution produce vesicles within an erupting lava. If the volume of these bubbles exceeds about 80% of the total volume, the magma will fragment explosively, which is the most dramatic (but not the only) way of creating a pyroclastic rock.

Before looking at **explosive eruptions** and their products, pause to consider the factors that would encourage an eruption to be explosive rather than effusive.

❑ Can you think of two factors likely to control this?

■ The main factors are the amount of volatiles available in the original magma, and the ease with which volatiles can escape as the magma approaches the surface.

The greater the percentage of volatiles a magma contains, the greater the volume of gas needing to escape as the magma approaches the surface. The more gas exsolving, the greater the likelihood that the rock will be fragmented, because if the gas cannot get out at a sufficient rate the volume of bubbles will reach the 80% threshold for explosive fragmentation. The ease of escape of this gas depends on the viscosity of the magma. If the viscosity is low, gas can escape quietly and efficiently by rise of bubbles to the surface and the volume of gas present at any instant may never reach 80%. However, if the viscosity is high, bubbles are inhibited from escaping, so the threshold is more easily reached. Therefore explosive eruptions are favoured by high viscosity and high volatile content of magma.

❑ Which chemical type of magma has the highest viscosity?

■ You learned in Section 4.2 that felsic magmas are the most viscous.

But what about volatile content? The most abundant volatile is water, so to a large extent we can answer this question by considering the water content of magmas.

Activity 6.3

You should now do Activity 6.3, in which you will make some observations bearing on this issue.

You should have concluded from Activity 6.3 that the magma types with the highest viscosity (felsic and intermediate) also tend to have a high water content. We should expect explosive eruptions to be particularly common in the case of felsic rocks, whose viscosity is highest.

❑ Which one of the igneous rocks you looked at in Activity 6.3 is pyroclastic?

■ The only pyroclastic sample is RS 25 (TS V).

This sample was introduced to you in Block 2 as an ignimbrite. Its composition is felsic, and we included it in your Home Kit to reflect the preponderance of explosive over effusive eruptions in volcanoes where felsic rocks are erupted.

Apart from the tendency of felsic magmas to fragment because of the explosive exsolution of gas, there is another reason why felsic eruptions are usually explosive. In Question 4.5, you saw a specific example of a hydrous melt of felsic composition that would be completely solidified before it reached a depth of 2 km (point C on Figure A4.2). Because most felsic magmas are hydrous, their ascent usually leads to their emplacement as plutons rather than to an eruption. However, there are exceptions. To appreciate why, you should now do Question 6.2.

Question 6.2

(a) Referring to Figure A4.2, suppose the magma at its source at 30 km was 100 °C hotter than the example in Question 4.5, but that it cooled at the same rate per km as it rose. At what depth would it be completely solidified?

(b) What if it started off under the conditions in Question 4.5 but lost heat less rapidly as it rose, at a rate of only 2 °C km^{-1}?

Thus, it is entirely feasible for water-saturated felsic magmas to get within a few hundred metres of the surface before solidifying. The pressure of gas trying to escape from such a shallow depth can trigger an explosive eruption (as a shower of solid fragments) *even though the magma would not have been able to reach the surface as a liquid*. Sudden depressurization of a magma chamber within a volcano because of partial collapse of the edifice or opening of a ring dyke can have a similar effect.

It is useful to distinguish explosive eruption deposits formed by material falling onto the ground (**airfall deposits**) from those formed by flow across the ground (**pyroclastic flows**). We will look at processes leading to airfall deposits first. A handy term to introduce at this stage is **volcanic ash** (usually abbreviated to just ash) which describes the fine fragments produced by the disintegration of magma. Its colour, black (if mafic) or grey (if intermediate or felsic), makes it resemble the ashes of a fire, but you should bear in mind that neither volcanic ash nor cinders (a term sometimes used for larger clinkery fragments) is a product of combustion. Ash is probably what forms the matrix between the blocks of vent agglomerate at Ardnamurchan that you saw in Activity 5.3.

6.3.1 Airfall-generating eruptions

The main factor controlling the nature of an explosive eruption is the viscosity of the magma. If viscosity is low, vesicles may coalesce as they rise and merge into large bubbles a metre or more across before reaching the top of the conduit. These arrive at irregular intervals and on bursting throw out a shower of ejecta. If the bubble bursts gently, the ejecta flops out around the rim of the conduit (Plate 6.6 and examples seen in Activity 6.1). More violent bubble bursting throws ejecta to greater heights (Plate 6.7), and this sort of episodic activity is described as **strombolian**, after the active Mediterranean volcano Stromboli.

Sometimes the expansion of vesicles in a basaltic conduit can cause magma and bubbles to accelerate upwards together, forcing them out of the vent at speeds of the order of 100 m s^{-1} as a fountain of incandescent lava, graphically described as a **fire fountain** (Plate 6.8). Most of the material thrown up in this way falls to the ground nearby, although some finer ash may be carried away by the wind. If the large chunks are still molten when they hit the ground, they can feed a lava flow. However, if they are sufficiently chilled in flight, they hit the ground as clinkery pieces called **scoria** (millimetres to centimetres in size). The larger pieces flung out by an explosive eruption are described as **volcanic bombs**. In the case of a scoria-producing fire fountain, the bombs chill less (because of their large size) and are typically soft and gooey so that they splat on the ground in a distinctive manner (Figure 6.9). Bombs thrown out more violently may take on aerodynamic shapes.

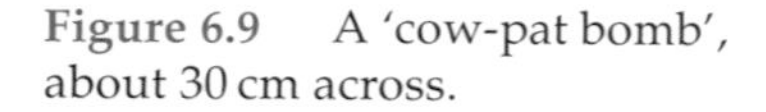
Figure 6.9 A 'cow-pat bomb', about 30 cm across.

Intermediate and felsic magmas are too viscous to allow vesicles to merge or to rise independently of the magma. Instead, vesicles remain small (< 1 cm). As exsolution proceeds, gas pressure forces the vesiculated magma up the conduit at high speed, emerging at the vent at several hundred metres per second. This usually gives rise to a type of eruption known as **plinian**, because the first detailed account of such an eruption (Vesuvius in AD 79) was written by Pliny the Younger. The eruption destroyed the cities of Pompeii and Herculaneum, and killed over 3500 people (one of whom was Pliny the Elder).

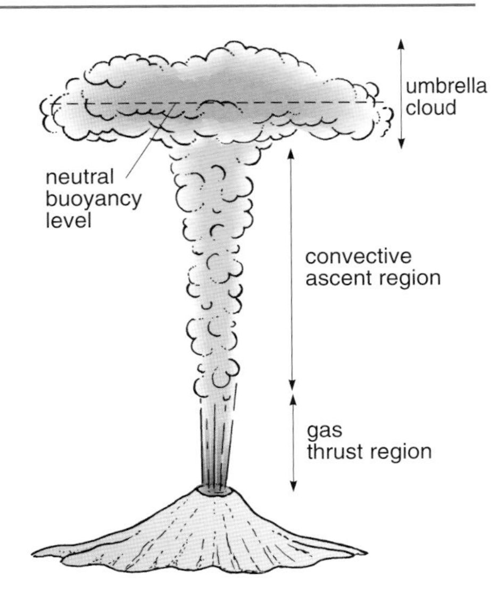

Figure 6.10 A plinian eruption column, showing the gas thrust region, the region of convective ascent, and the umbrella cloud.

A plinian eruption may last for hours or days, and is marked by a grey column of ash known as an **eruption column** rising to kilometres or even tens of kilometres above the vent. The physical processes occurring within a plinian eruption column make it convenient to describe it in three parts (Figures 6.10 and 6.11). In the lower part of the column, fragments are driven upwards by the force of expanding gas in the conduit. This is called the gas thrust region, and is equivalent to the entire eruption column of a fire fountain. However, in a plinian eruption column a lot of air is drawn in and heated by contact with the hot ejecta, causing the air to expand. Mixed with the hot volcanic gases (and despite the weight of the ash particles within it), this is buoyant relative to the cold surrounding air. It rises convectively above the gas thrust region and is called the convective ascent region of the eruption column. Eventually, the column reaches a height where it is neutrally buoyant, and spreads out to form an umbrella cloud. Because the umbrella cloud is not rising, all the ash is now able to fall out, and the extent of the airfall deposit therefore depends on how far the umbrella cloud spreads.

Figure 6.11 A plinian eruption column rising 20 km above Lascar volcano, in the Andes of northern Chile. The convective ascent region and the umbrella cloud are well seen, but collapses of the outer part of the column hide the gas thrust region from sight.

A plinian eruption creates an airfall deposit with clearly recognizable characteristics. The larger ejecta falls too rapidly to be carried by the convective ascent region, so it hits the ground relatively close to the vent, although some bombs can travel several kilometres. Slightly smaller material falls out of the sides of the convective ascent region, and is dispersed further than the majority of the bombs. Ash reaching the umbrella cloud falls at a speed that decreases with decreasing particle size, so the finest material settles out last and is dispersed furthest from the volcano.

This is an appropriate time to introduce the term **tephra** (from the Greek word for 'ash') which refers to airfall material in general; hence the term tephrachronology applied to the use of widespread airfall deposits that have been radiometrically dated as stratigraphic marker horizons. Table 6.1 shows the accepted grain size terminology for tephra.

Table 6.1 Terminology for tephra according to grain size (i.e. particle size).

Grain size/mm	Pyroclastic fragments (tephra)
>256	coarse bombs and blocks
64–256	fine bombs and blocks
2–64	lapilli
$\frac{1}{16} - 2$	coarse ash
$< \frac{1}{16}$	fine ash

Particularly in felsic eruptions, many of the fragments in the lapilli size range are lumps of glassy froth described as **pumice**. Most ash particles are sharp glassy fragments representing the walls of bubbles, though some consist of isolated phenocrysts or are fragments of phenocrysts.

Not only does coarser ejecta fall closer to the vent, but the total amount falling from the umbrella cloud decreases outwards. Plinian airfall deposits thus have a tendency to become both thinner and finer-grained away from the source. However, there is a very important external factor that has a considerable influence on the shape of this pattern.

❑ Can you think what this is?

■ The dispersal of the umbrella cloud is controlled by the *wind*.

When there is a wind, the umbrella region is blown off to one side (Plate 6.9) so tephra is deposited asymmetrically. When the thickness of an airfall deposit is contoured, the pattern is consistent with the wind direction during the eruption. Try Question 6.3 to ensure that you understand this.

Question 6.3 Look at Figure 6.12. Explain from which direction the wind was blowing during this eruption. (*One or two sentences*)

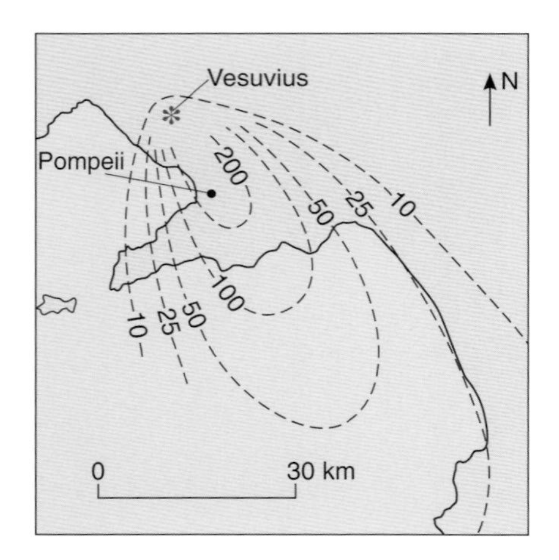

Figure 6.12 Contours of equal thickness (isopachs) of the airfall deposit from the AD 79 plinian eruption of Vesuvius (for use with Question 6.3). Thicknesses are in centimetres.

Other styles of explosive eruption are determined by the relationship between volatiles and magma. Sometimes a volcanic conduit is choked by debris from earlier eruptions, and this is cleared out by explosive release of gas from an underlying magma chamber. Little or no fresh magma is erupted, and the tephra consists largely of redistributed fragments of older volcanic rocks. This style of eruption is known as **vulcanian**, after the Mediterranean island volcano of Vulcano, whose most recent large eruptions exemplify this style of activity. Alternatively, a violent explosion can be triggered when magma suddenly encounters water. This can simply be groundwater, or be caused by eruption below less than a few hundred metres of water, or when lava flows into a sea or lake. This kind of explosive eruption is described as **phreatic***, and is driven by the sudden conversion of water to steam.

* In reading beyond this Course, you may encounter a distinction between 'phreatic' applied only to explosions of steam, mud and fragments of old volcanic rock, and 'phreatomagmatic' or 'hydrovolcanic' applied to explosions of steam with fresh magma. Here we use 'phreatic' to encompass the whole spectrum.

Figure 6.13 Cross-section showing a 25 cm-thick airfall deposit, decreasing upwards in grain size from several mm at the base to less than 1 mm near the top. The coarse base of the next airfall deposit is visible above.

Generally speaking, plinian, vulcanian and phreatic eruptions fall across a scale of decreasing duration and column height. However, they share common eruption column characteristics and the dispersal of their deposits follows a similar pattern, with the result that the nature of an ancient eruption can best be ascertained by considering the nature and condition of the tephra.

A notable characteristic of explosive eruptions is that they usually wane in intensity as they progress. This means that not only does the average size of tephra decrease away from the vent, it also decreases upwards within the deposit (Figure 6.13), although the trend may be interrupted by temporary bouts of increased eruptive vigour. The interpretation of airfall deposits is made complicated not only because of variations in the type and size of the clasts, but also because such deposits may be interbedded with material deposited from pyroclastic flows, which we will look at in the next Section.

6.3.2 Pyroclastic flows

In Block 4, you will study the lateral transport of particulate material to produce deposits of clastic material. Here we will confine our attention to a descriptive account of how pyroclastic material may be transported along the ground. When this happens, ash, pumice and other fragments are carried along in a flow of air, with clasts bouncing frequently off one another during motion. The flow as a whole behaves as a dense fluid (it is said to be in a fluidized condition), and the behaviour of the agitated clasts can be likened to that of individual molecules in a liquid. A simple cross-section through a **pyroclastic flow** is given in Figure 6.14 (overleaf). The head of the flow is particularly low in density because of the entrainment of air, and is sometimes described as a **surge**. The flow process here can result in traces of ripples within the resulting deposit, similar to those seen in sandstone deposited by flowing water. The finest material tends to blow away as an ash cloud that rises from the head or body of the flow. Most deposition occurs from the body of the flow, and the resulting deposit is massive and lacks distinctive structure. Movement ceases when the flow loses energy and the air escapes.

There are several ways in which a pyroclastic flow can be generated. One is **column collapse**, when a plinian eruption column (or maybe just its outer part) becomes unstable and the tephra within it topples to the ground instead of rising upwards. For a plinian eruption column to be stable, the rate of magma discharge and the radius of the vent (which are related) must stay below a

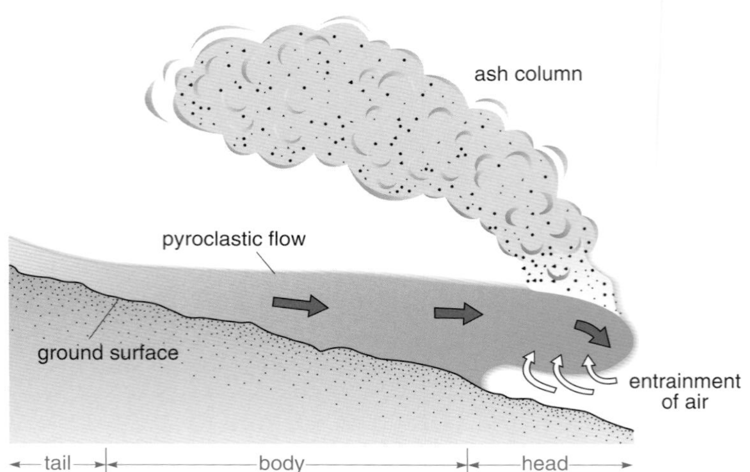

Figure 6.14 Cross-section through a pyroclastic flow (typically hundreds of metres to tens of kilometres in length), which is moving downslope from left to right. (See also Plate 6.8b in Block 2.)

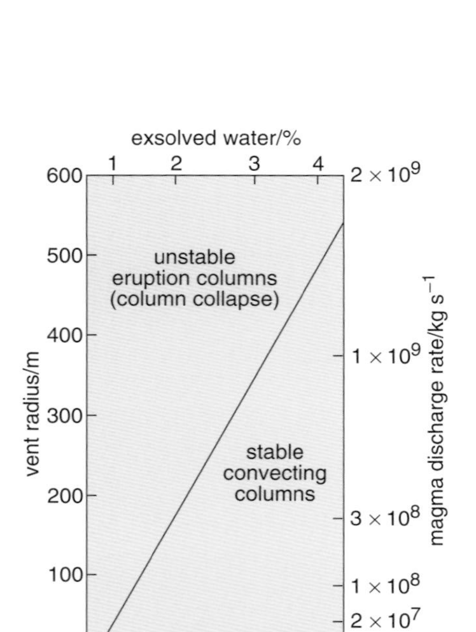

Figure 6.15 The stability field for convecting plinian eruption columns and the field corresponding to column collapse. Note that exit velocity is linked to the percentage of exsolved water, and magma discharge rate is linked to vent radius.

certain threshold, whereas the exit velocity at the vent and the volatile content of the magma (which also tend to be related) must remain high enough. The relationship is shown graphically in Figure 6.15. Study this Figure, and then attempt Question 6.4.

Question 6.4

(a) It is reasonable that the vent can grow wider during the course of a plinian eruption, because of blasting away or collapse of its rim. Suppose ejecta was being flung upwards from a vent 150 m in radius at 300 m s^{-1}. Explain what would happen if the vent became widened to 200 m radius (without any other factor changing).

(b) Suppose instead that from the initial conditions in (a), the amount of exsolved water in the magma fell to 1% (a likely occurrence if volatiles are escaping from the source at a proportionally faster rate than magma). What would happen in this case?

Question 6.4 should have suggested that column collapse is a common occurrence in the later stages of a plinian eruption. Although this happened in the archetypal AD 79 plinian eruption of Vesuvius, it is not inevitable, which is fortunate because the pyroclastic flows generated by column collapse are among the most devastating volcanic hazards.

When material from a collapsed column hits the ground, it is deflected sideways into a pyroclastic flow moving at several hundred metres per second that is capable of travelling for 100 km or more. The resulting deposit is an ignimbrite (Block 2 Section 6.4) the main part of which consists of ash interspersed with larger clasts. These may be two types: pumice and **lithic fragments**, the latter being pieces of dense (non-vesicular) rock. The pumice is formed in the eruption, and the lithic fragments are usually pre-existing rock ripped off the walls of the conduit. Usually, the flow process concentrates the larger clasts towards the top of the flow (a phenomenon called *reverse grading*) but only pumice fragments, being lighter, tend to reach the top. Overlying the ignimbrite may be a fine ash deposit, representing airfall either from the ash column that rose from the flow itself or from a plinian eruption column. A generalized cross-section through a complete ignimbrite is shown in Figure 6.16.

Often, the ash and coarser pumice fragments in an ignimbrite flow are still hot when they come to rest, and may become welded together. This is particularly likely near the base of the flow where the process is aided by compaction under the weight of the upper part of the deposit. The pumice fragments are squashed while they are still hot and pliable into forms described as fiamme (pronounced 'fee-am-eh'), which is Italian for flame (Plate 6.10), whereas the denser lithic fragments are unaffected.

The other common kind of pyroclastic flow is a **block-and-ash flow**, which is a pyroclastic flow in which the larger clasts are dense blocks (usually of andesite or dacite) rather than pumice. A block-and-ash flow can be triggered by non-explosive collapse from the face of an extruded dome of andesitic or dacitic lava (a **lava dome**) as in the case of the flow shown in Plate 6.8b of Block 2, or be caused by an explosive event that flings out sufficient quantities of blocks (perhaps during a plinian eruption). The flow process is much the same as in an ignimbrite, from which the resulting deposit can be distinguished chiefly by its lack of pumice. Sometimes the interior of the collapsed dome is so hot that the moving flow is incandescent and is described as a nuée ardente (pronounced 'noo-eh ardont'). The name, meaning 'glowing cloud' in French, arises from the eruption of Mount Pelée that killed 29 000 people and destroyed the large town of St Pierre on the French Caribbean island of Martinique in 1902 (Figure 6.17).

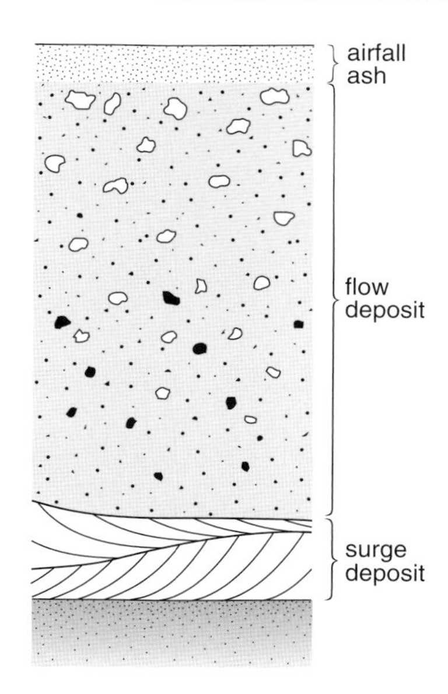

Figure 6.16 Generalized cross-section through an ignimbrite. Pumice fragments are shown white, and lithic fragments black. Total thickness may be anything from a metre to hundreds of metres.

Question 6.5 From what you have learned about their dispersal mechanisms, how would you expect the distribution and thickness of pyroclastic flows and airfall deposits to differ, particularly in respect to the local topography? (*Two or three sentences*)

Although the answer given to this question is basically correct, there are some important secondary considerations. Pyroclastic flows travel so fast that their momentum can carry them up opposite-facing slopes, so although flow deposits tend to be thickest in valley bottoms they are not entirely confined by topography. On the other hand, when airfall lies on steep slopes, it is particularly vulnerable to being washed away during rainstorms, so it may not be preserved in such places but will be redistributed in the form of a volcanic mudflow known as a **lahar**.

A previously unrecognized mechanism for sideways transport of pyroclastic material became apparent when Mount St Helens (Washington State, USA) erupted in May 1980. This eruption was triggered by the injection of dacitic

Figure 6.17 St Pierre, Martinique, a week after the nuée ardente eruption of May 1902 from Mount Pelée. The town was completely destroyed, although the deposit amounted to only a few centimetres of ash. The summit of Mount Pelée is in cloud in the background.

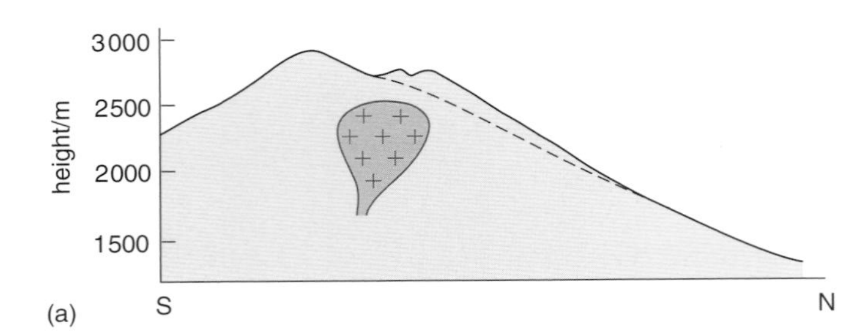

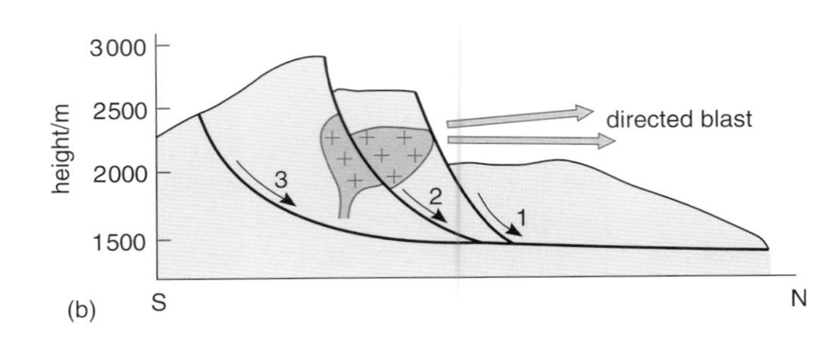

Figure 6.18 Cross-sections through Mount St Helens (no vertical exaggeration). (a) Morning of 18 May 1980 showing the position of the high-level dacite magma chamber and the swelling of the northern flank of the volcano relative to its previous profile (dashed line). (b) A series of three landslips that generated the debris avalanche deposits, the first of which exposed the dacite magma chamber allowing it to de-gas explosively, causing the directed blast.

magma into an off-centre chamber within the volcano (Figure 6.18a). The northern side of the volcano swelled upwards by as much as 100 m until this flank became unstable and began to slip. This undermined the summit, which collapsed in turn so that within half a minute more than five cubic kilometres of material was sliding northwards as a debris avalanche. Collapsing volcanoes are a phenomenon to which we shall return in the next Section. What concerns us here is that this sudden collapse removed the confining pressure on the magma chamber, so that the volatiles within it exsolved explosively. This shot fragmented dacite sideways at supersonic speeds (>300 m s^{-1}) in a **directed blast** (Figure 6.18b) that overtook the slower-moving debris avalanche and flattened trees over an area of nearly 600 km^2 (Figure 6.19).

Figure 6.19 These mature Douglas Fir trees were knocked over and stripped of their bark by the May 1980 directed blast from Mount St Helens, which is visible (lacking its previous summit) in the background. The deposit consists mostly of fine ash only a few cm thick, with a few chunks of pumice.

6.4 Volcanoes

We now consider how the individual processes of volcanic eruption combine to build volcanoes. A volcano's shape depends primarily on the types and relative amounts of effusive and explosive activity that went into its construction. In the description that follows, we sometimes refer to basaltic (meaning mafic), intermediate and felsic volcanoes. This is a convenient simplification, indicating the type of magma most commonly erupted. It is important that you do not form the false impression that a volcano only ever erupts magma of a single composition. First, it is common for the silica content of magma to vary by 10% or more during a single eruption (which indicates that the magma chamber was compositionally zoned prior to the eruption). Secondly, the average composition may evolve during a volcano's lifetime of tens of thousands to millions of years. Successive eruptions often become increasingly silica-rich as activity wanes (because magma supply is less and the time available for fractionation is greater).

Plate 5.1 Photograph looking along an eroded dyke of early Tertiary age in west Texas. Erosion has removed the soft sediments on either side of the dyke preferentially, so that the eroded top of the dyke stands above the general level of the terrain. This dyke is one of a radial swarm, associated with an igneous complex forming the hill in the distance.

Plate 5.2 Vertical columnar joints in the Whin Sill, formed by thermal contraction as the sill cooled. This view shows a 20 m high quarry face. Neither the top nor the bottom of the sill is visible.

Plate 5.3 Thin section through the base of the Whin Sill (between crossed polars), showing an area about 1.4 mm long. In region A, crystals of pyroxene and plagioclase feldspar occur in a finer-grained and partly glassy matrix. Region B is the actual chilled margin which is glassy throughout, except for a couple of olivine crystals which presumably grew within the magma before it was emplaced. Regions C and D are within the country rock, which is limestone in this case. You should recognize the pastel interference colours of the calcite crystals in D. In region C (only about 0.2 mm wide), some new metamorphic minerals have begun to grow.

Plate 6.1 An a'a flow surface (left) partially overridden by a subsequent ropey pahoehoe flow (right), Hawaii. The flows are identical in composition, but the pahoehoe looks paler because of its shiny fresh glassy surface.

Plate 6.2 Lineated sheet flow or 'toothpaste' pahoehoe (Craters of the Moon lava field, Idaho).

Plate 6.3 A ropey pahoehoe lava surface in the process of formation, Hawaii. Where the surface crust has been stretched, the incandescent molten interior is revealed, but where the crust is thicker (though still pliable), it is becoming pleated over because of drag on its underside by the more rapidly moving molten interior. The bystander's feet are resting on toey pahoehoe.

Plate 6.4 An a'a flow in motion (Hawaii). (*US Geological Survey*)

Plate 6.5 View looking down on a blocky andesite lava flow from near its source, showing a prominent levée on either side of the main channel, and the lobate outline of the more distant part of the flow. The total length of this flow is about 5 km (Momotombo volcano, Nicaragua; 1905 flow photographed in 1989).

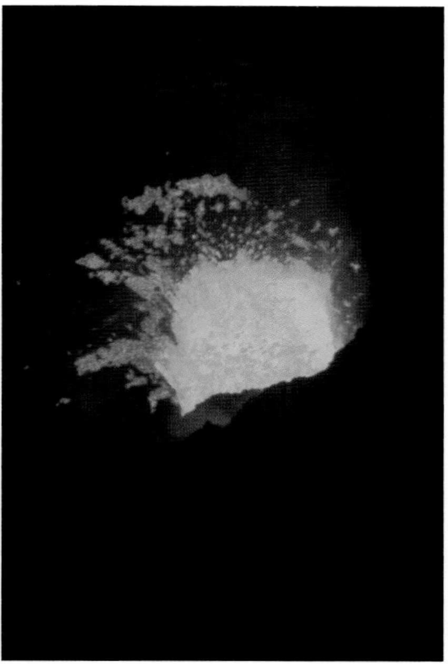

Plate 6.6 Photograph taken at night a few seconds after a mild strombolian bubble-bursting event at the top of an active conduit in Masaya volcano, Nicaragua. The vent is about 5 m across and is nearly brim-full of molten lava. The bubble that has just burst has thrown out molten spatter, which is already beginning to cool.

Plate 6.7 This cloud of ash was flung out by a strombolian eruption from a vent on the floor of the crater (Stromboli, Italy).

Plate 6.8 A 200 m-high fire fountain on Mount Etna. This fountaining event continued for more than an hour.

Plate 6.9 View of the plinian eruption column from the May 1980 eruption of Mount St Helens. The wind was blowing from right to left, and slightly away from the viewpoint. (*US Geological Survey*)

Plate 6.10 Squashed fragments of pumice, or fiamme, in an ignimbrite. The largest in this example is about 15 cm long. There are only a few small lithic fragments here, and these are not squashed.

Plate 6.11 Two scoria cones on a rift on the south-east flank of Mount Etna. The asymmetry of their rims is caused by the wind during eruption blowing from left to right and partly away from the viewpoint, thereby carrying most ejecta towards the far right.

Plate 6.12 A view from space showing Socompa volcano and its debris avalanche deposit. The volcano is at the lower right (cf. Figure 6.26). New lava flows and domes have begun to fill the amphitheatre formed by the collapse event. Image is 40 km across. (*Peter Francis, Open University*)

Plate 6.13 A flat, relatively featureless ignimbrite plateau in the Andes, with unrelated stratocone volcanoes in the background.

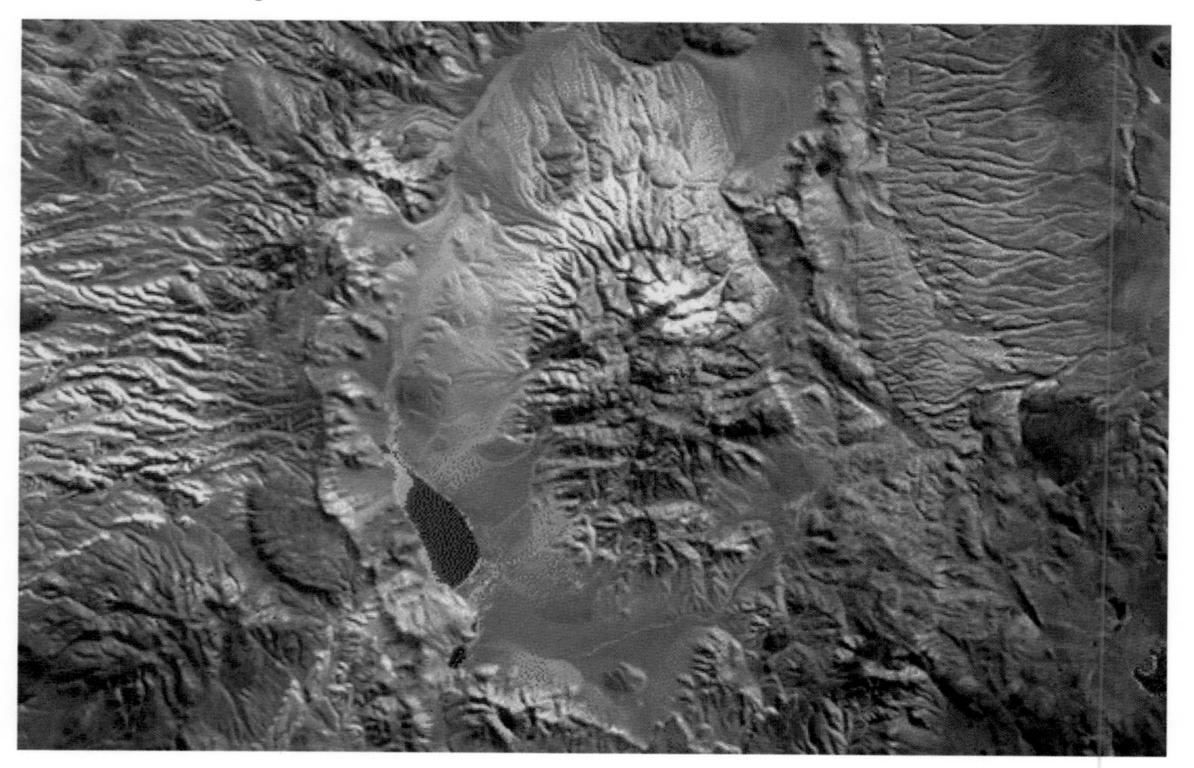

Plate 6.14 View from space showing the oval-shaped caldera of Cerro Galan, Argentina, which measures about 45 km north–south and 25 km east–west. The caldera was formed by an explosive eruption about 2.2 million years ago. Part of the resulting ignimbrite plateau is visible on either side, and the middle is occupied by a resurgent centre. Some post-caldera dacite flows sit above the ring fracture at the northern (top) edge of this view. The blue feature in the south-west of the caldera is a lake. (*Peter Francis, Open University*)

Plate 8.1 Garnet crystals (high relief) partly replaced by pale green chlorite as a result of retrograde metamorphism. (Plane-polarized light; width of image = 1 mm.)

Plate 8.2 Fault breccia: fragments of feldspar lie within a matrix of pulverized quartz.

Plate 8.3 (a) Mylonite from the Moine Thrust, Scotland. (b) Ribbon texture from recrystallized quartz in a mylonite (seen between crossed polars; width of image = 5.5 mm).

(a)

(b)

(a)

Plate 8.4 (a) Pseudotachylite (dark band) formed within an amphibolite gneiss from an extensional fault zone in the Alps. Coin diameter is 2.5 cm. (*Wolfgang Müller, ETH, Zurich*)
(b) Pseudotachylite resulting from meteorite impact in Vredefort, South Africa. The dark bands in the quarry face are pseudotachylite, and below these are disrupted fragments of rock that have collapsed downwards through the melt. (*John Spray, University of New Brunswick*)

(b)

Plate 9.1 Folds in bedded greywackes from Shap, Cumbria.

Plate 10.1 Folding and thrusting at Broad Haven, south-west Wales. Looking east. J.L. Roberts (1989) *The Macmillan Field Guide to Geological Structures*, Macmillan.

Plate 11.1 (a) Muscovite flake buckled by deformation of muscovite schist. (Plane-polarized light; width of image = 2.2 mm.) (b) The same grain seen between crossed polars. (*Christophe Prince, Open University*)

(a)

(b)

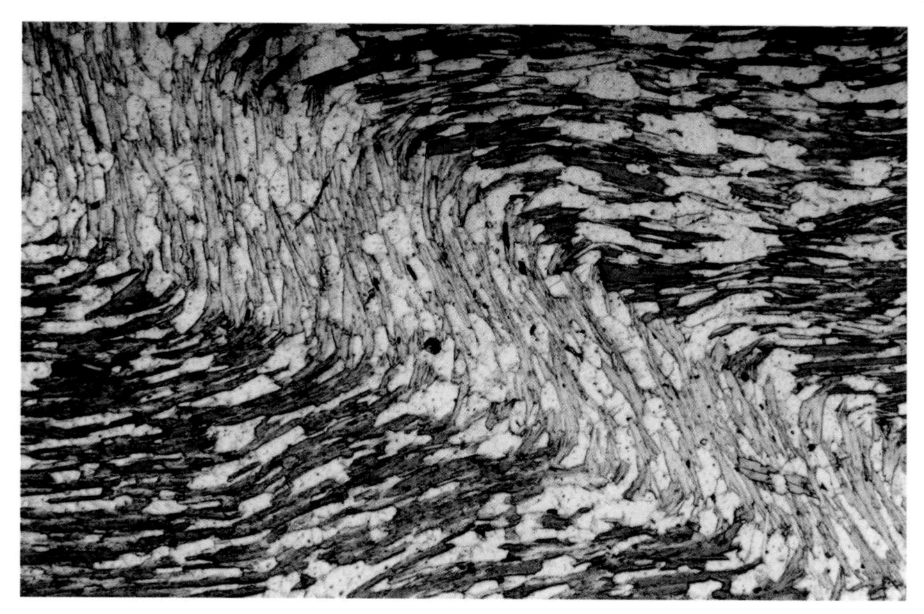

(a)

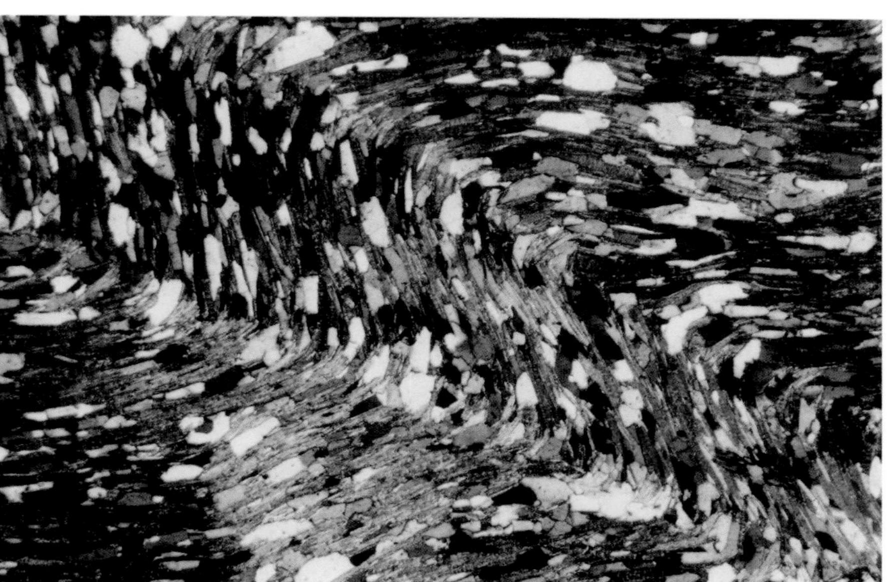

(b)

Plate 11.2 An originally planar schistosity defined by layers of biotite and quartz, buckled into a fold by a second deformation episode. (a) Plane-polarized light; width of image = 3.5 mm; (b) seen between crossed polars. (*Christophe Prince, Open University*)

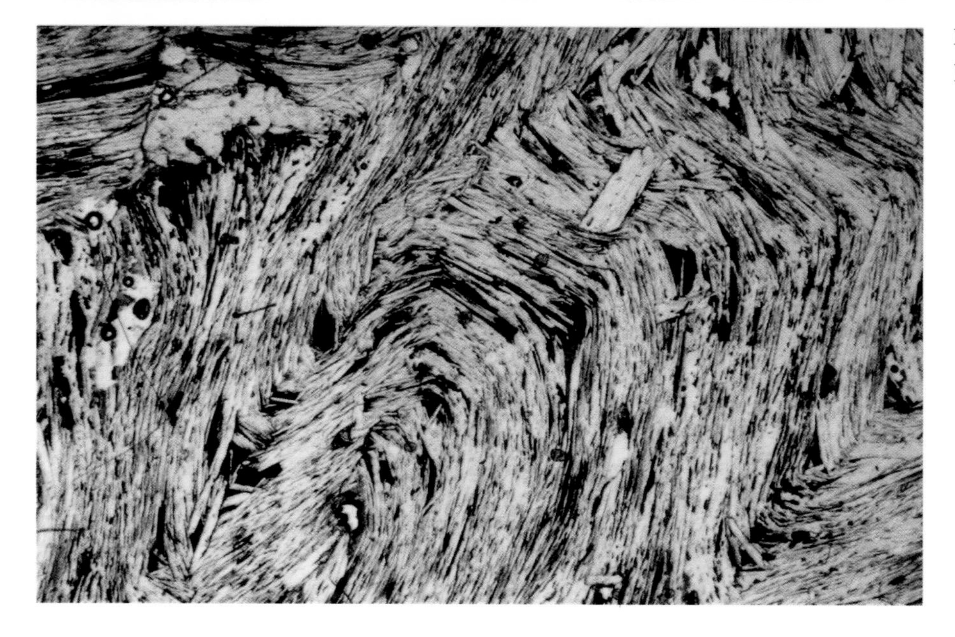

Plate 11.3 Muscovite schist (plane-polarized light, width of image = 3.5 mm).

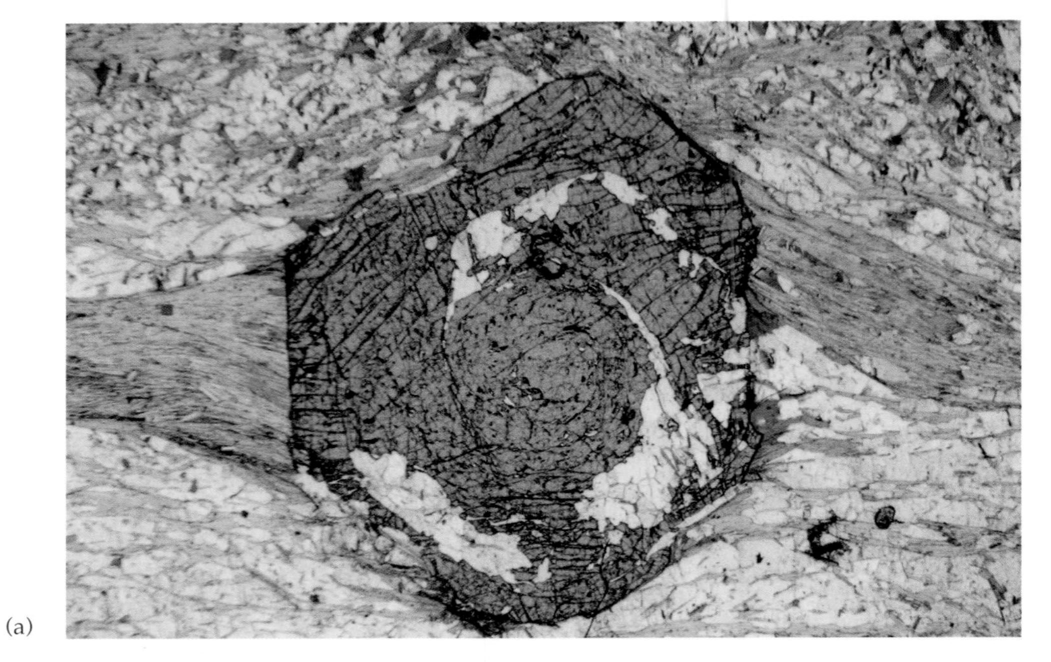

(a)

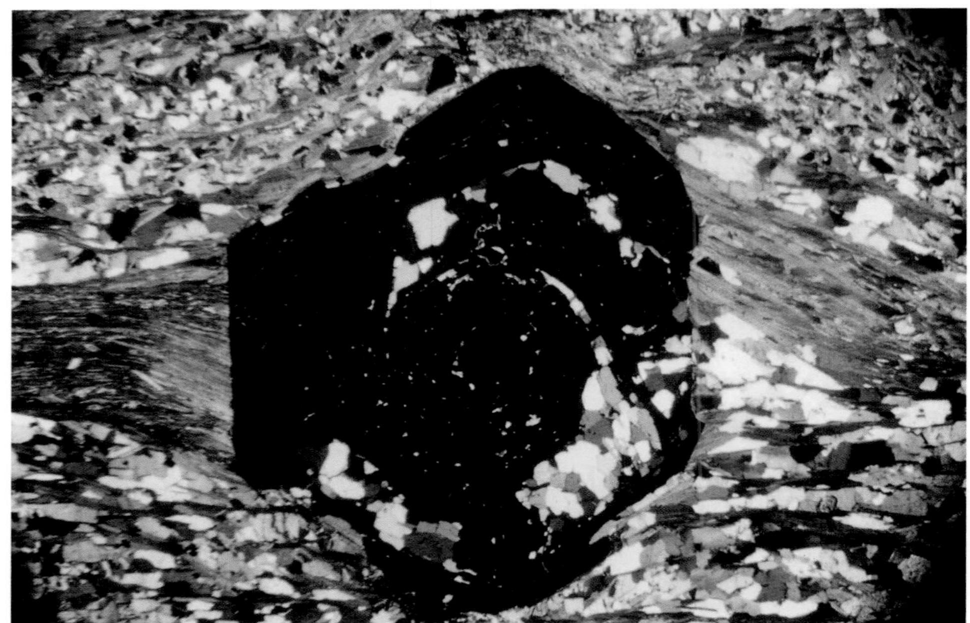

(b)

Plate 11.4 (a) Garnet-biotite schist. (Plane polarized light; width of image = 1 mm.) (b) Viewed between crossed polars.

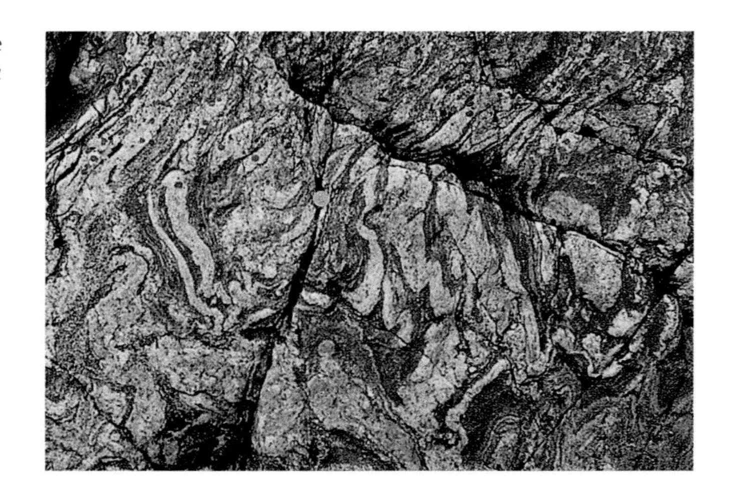

Plate 12.1 Migmatite gneiss from the inner zone of the Caledonian orogenic belt in Scotland, showing complex folding and evidence for melting. J.L. Roberts (1989) *The Macmillan Field Guide to Geological Structures*, Macmillan.

6.4.1 SHIELD VOLCANOES

- ❑ You have read that basaltic volcanism is mostly effusive. Bearing in mind the viscosity of basalt lavas, would you expect basaltic volcanoes to have steep or gentle slopes?
- ■ Basaltic lava has very low viscosity (Section 4.2), and so it forms thin flows capable of spreading a long way from their source. Basaltic volcanoes should therefore have gentle slopes.

This is indeed the case, as exemplified by the basaltic volcanoes forming Hawaii's Big Island (Figure 6.20) where a gentle convex profile has been created by a series of eruptions fed mainly from the volcano's central vent. This shape is described as a **shield volcano**, because of its resemblance to a warrior's shield laid on its side.

Figure 6.20 Oblique view from an aircraft showing the snow-capped summit of Mauna Kea, Hawaii, and its darker lower slopes rising above the cloud deck. Mauna Kea and its neighbour Mauna Loa are the largest shield volcanoes on Earth; their summits are 8 km above their base, which is 4 km below sea-level and has a diameter of nearly 200 km.

Most volcanoes, including shields, have a roughly circular depression at the summit containing the vent. By convention this is described as a **crater** if less than 1 km in diameter and as a **caldera** if bigger than this. A crater or caldera is usually considerably wider than the vent and the conduit that feeds it, and is formed by a mixture of collapse and subsidence on ring faults. If the vent or magma chamber have changed location during the lifetime of the volcano, then the summit area is likely to consist of a series of overlapping craters or calderas (Figure 6.21).

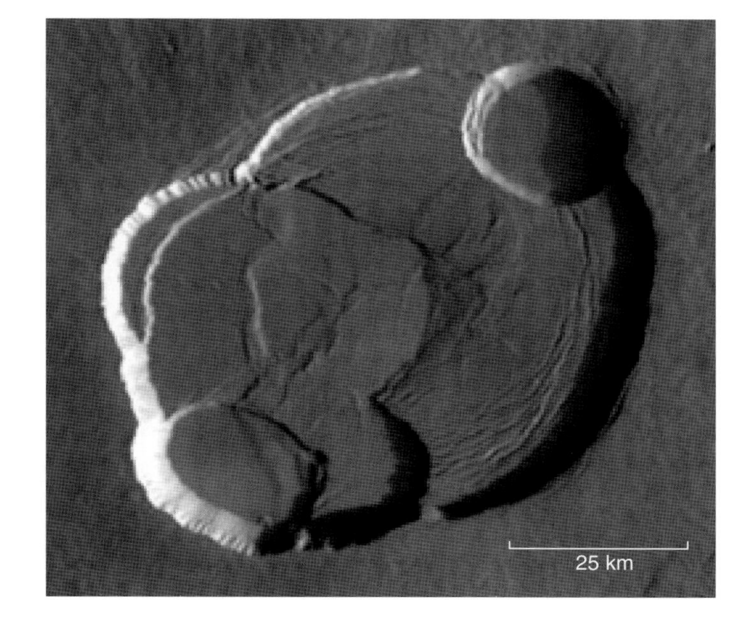

Figure 6.21 Overlapping calderas at the summit of a basaltic shield volcano. This example is from Olympus Mons on Mars, which is a larger equivalent of Hawaiian-type volcanoes. The flat floor of each caldera is thought to be lava that partly filled it as a lava lake.

6.4.2 Stratocone volcanoes

When magma of generally intermediate composition erupts to form a volcano, a strikingly different shape results, thanks to a combination of two factors. First, intermediate lavas are more viscous than mafic lavas, so they flow less far from their source. Secondly, explosive eruptions are more common at intermediate volcanoes than at basaltic ones. The resulting mixture of short stubby lava flows, pyroclastic flows and airfall deposits builds a volcano that is a relatively steep-sided cone, like the classic Mount Fuji in Japan or the Andean example in Figure 6.22. The alternation of pyroclastic and lava layers (Figure 6.23) has led to the name **stratocone** being applied to this type of volcano. Stratocones are commonly between 10 and 40 km in diameter and 1–4 km high, and may have summit craters or calderas up to a few km in diameter.

Figure 6.22 A stratocone volcano (Acamarachi) in the Andes near the Chile–Argentina border.

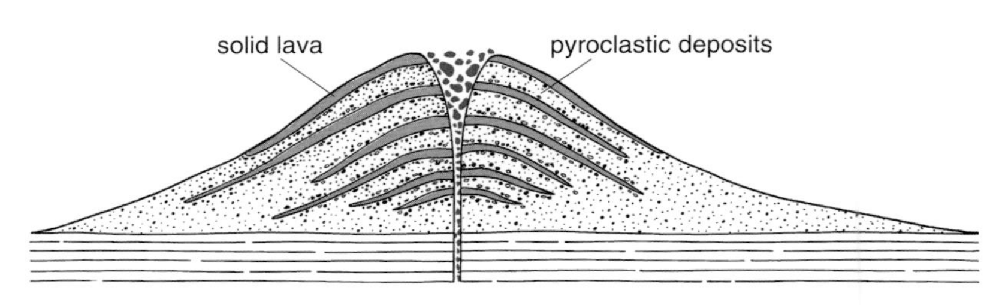

Figure 6.23 Diagrammatic cross-section through a stratocone volcano (not to scale).

Although some stratocone volcanoes end their days peacefully, becoming quietly extinct after a few hundred of thousands of years of activity, others collapse one or more times like Mount St Helens (Figure 6.18) or go out with a bang in the form of a large-volume caldera-forming felsic eruption. Figure 6.24 illustrates a classic example.

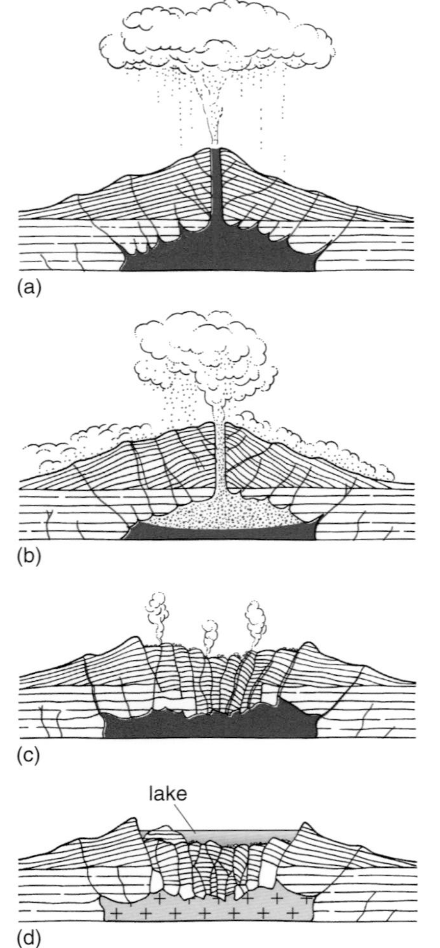

Figure 6.24 (in margin) The origin of the caldera at Crater Lake, Oregon. (a) The ancestral stratocone volcano, erupting mainly intermediate magma since about 400 000 years ago. (b) Cataclysmic eruption of felsic magma about 7000 years ago, producing a large volume of plinian airfall and ignimbrites, and followed immediately by (c) collapse into the partly evacuated magma chamber forming a 8 km-diameter caldera. (d) The situation today. There have been some minor andesite and dacite eruptions since the caldera was formed. The magma chamber is largely solidified and the caldera is occupied by a lake of water.

6.4.3 VOLCANIC RIFTS

Shield volcanoes and stratocones are usually portrayed as symmetrical features. However, many such volcanoes are elongated, because some of the magma reaches the surface along a fissure rather than up the main central conduit. The fissure may result from regional tectonic extension, but more usually it appears to be related to instability in the volcanic edifice, causing it to fracture. Examples are shown in Figure 6.25 where it is apparent from the map that many of the flows on Mauna Loa have been fed from volcanic fissures (described as *rifts*) running north-east and south-west from the summit, with the result that the shield of Mauna Loa is elongated in this direction. The smaller (and younger) volcano Kilauea is even more elongated. Its south-west to north-east rift system can be explained by a tendency of the volcanic edifice to spread towards its unbuttressed south-east side.

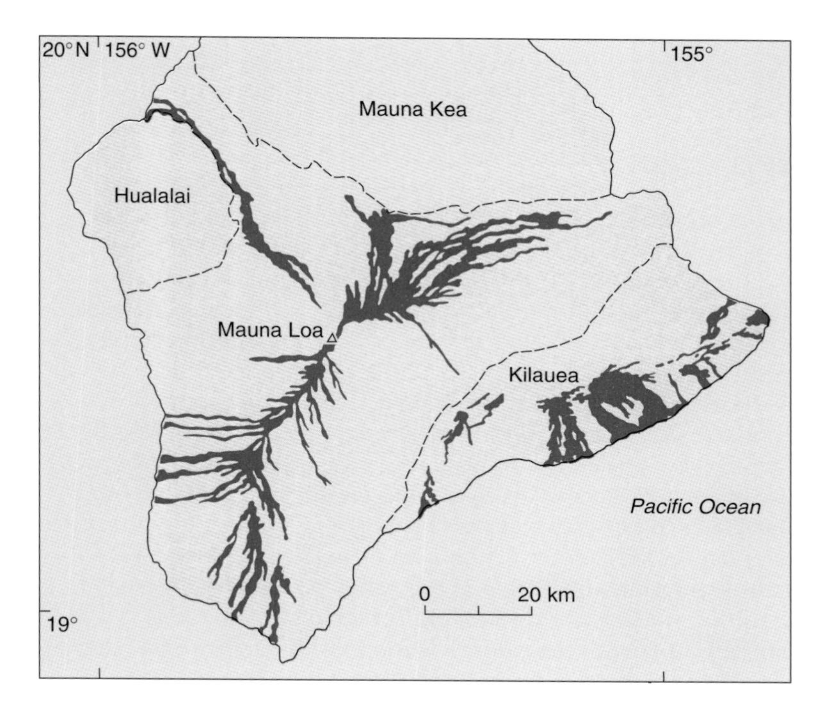

Figure 6.25 Map of Hawaii showing lava flows erupted between 1843 and 1984 (on Mauna Loa) and between 1750 and 1998 (on Kilauea). The preponderance of sources along rifts trending north-east and south-west from either summit is obvious. The names are those of the individual basaltic shields making up the island.

- ❑ When a volcano is cut by fissures, these offer an alternative route for magma to the reach the surface other than at the main vent. Can you suggest why magma would tend to follow this route if available?
- ■ Fissures must intersect the surface on the flanks of the volcano at a lower altitude than the central vent, so if a fissure is open less pressure is required to bring magma to the surface this way.

In practice, fissures are often clogged by a solidified dyke that fed previous activity. Thus, eruptions tend to occur through the summit vent if fissures are blocked, but somewhere along the rift if a fracture has reopened in response to underlying tectonic extension or spreading of the edifice.

When a fissure eruption occurs on a volcano's flanks, it may in extreme cases begin by fire fountaining along several kilometres of the rift, as you saw in Activity 6.1. However, the eruption soon becomes focused at a single point (presumably because other pathways become blocked by magma chilling against the walls of the fissure), where a cone of scoria, described as a **scoria cone** or cinder cone (Plate 6.11), builds up and which is the source of most of the lava flows.

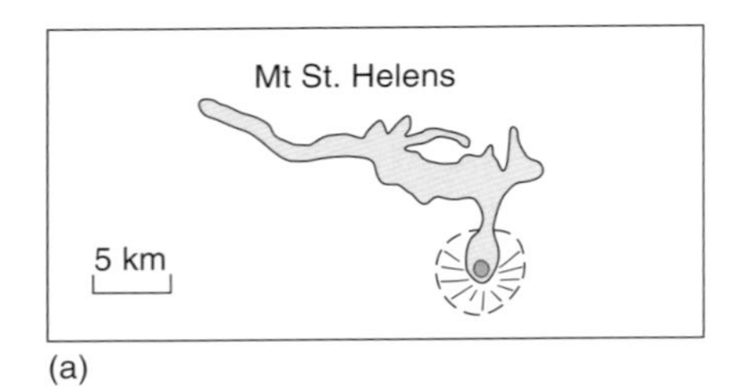

(a)

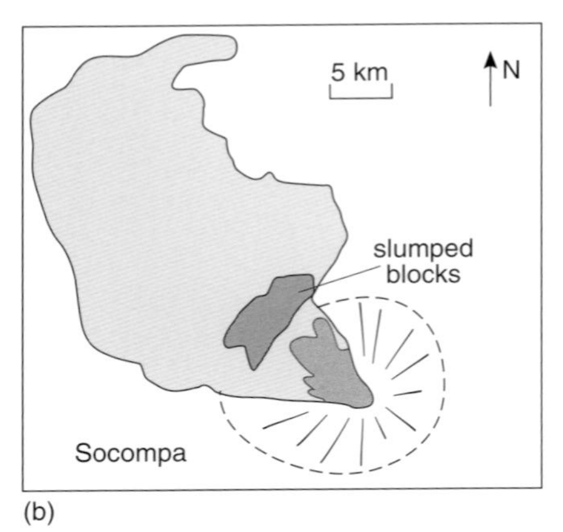

(b)

Figure 6.26 Map comparing the sizes of the debris avalanche deposits (shaded) at Mount St Helens (1980) and the much larger example at Socompa volcano, Chile (72 000 years ago). An area of slumped blocks each >1 km in size is shown in the mouth of the Socompa collapse. Lava extruded since each collapse is indicated within the amphitheatre left by the collapse in each case.

6.4.4 COLLAPSING VOLCANOES

Any mountain is inherently unstable, and unless it is strong and internally coherent gravity will cause it to collapse or spread sideways. A volcano is particularly vulnerable, because the disruption to its structure associated with the injection of dykes and other magma bodies makes it increasingly prone to **volcanic collapse**. You have already met the devastating consequences of volcanic collapse in the case of Mount St Helens, whose 1980 eruption brought this previously unrecognized phenomenon to world attention. Prehistoric debris avalanche deposits have now been discovered around many other volcanoes (Figure 6.26, Plate 6.12), at some of which there has since been time for the collapsed sector to have become filled with new erupted products. Not all volcanic collapses are thought to have led to Mount St Helens-style directed blasts. Moreover, many have occurred underwater and very extensive debris avalanche deposits have been mapped on the Pacific floor around the Hawaiian Islands.

6.4.5 IGNIMBRITE CALDERAS

Where felsic magma is erupted in abundance, the resulting landform is quite unlike a picturebook volcano. In the rare instances where eruption of felsic magma is extrusive, the result is thick flows like those pictured in Figure 6.6. However, as you have seen, the high viscosity and high water content of felsic magmas means that most felsic eruptions are explosive. The largest volume felsic eruptions produce ignimbrites, which are either erupted from fissures (that can be hard to locate afterwards) or associated with the production of calderas about 20 km across. Figure 5.7b indicated how a ring dyke would be emplaced during subsidence of the roof of a shallow magma chamber. Magma reaching the surface in such an event, which can empty about half the magma from the chamber, is likely to do so explosively. This will produce an ignimbrite flow that can travel over 100 km, and cover an area of several thousand km^2 described as an **ignimbrite plateau** (Plate 6.13). Doming caused by magma injection prior to eruption means that the caldera rim is higher than the surrounding terrain, but the rest of the caldera is more in the nature of a hole in the ground than a mountain. After an ignimbrite-forming eruption, lava flows or domes are usually erupted within the caldera (often fed by reactivation of the ring dykes, as hinted at by the domes drawn in Figure 5.7b) over a period that may last several hundred thousand years. If magma is reinjected into the magma chamber, the floor of the caldera will be domed upwards to form a broad rise within the caldera that is described as a **resurgent centre**, and which may eventually overtop the rim. These features can all be distinguished on Plate 6.14.

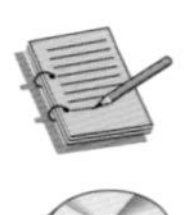

Activity 6.4

This has been a long Section, in which you have encountered a wide variety of processes at many different scales. It can be difficult to understand volcanic eruptions from the printed page alone, so we have compiled some video footage of eruptions in the video sequence *Eruption styles* on DVD 2, which you should now watch as Activity 6.4.

6.5 Summary of Section 6

- The surface textures of lava flows depend on viscosity, rate of flow and the efficiency of cooling. These factors also affect the shape of a lava flow field.
- Explosive eruptions occur when decreasing pressure allows volatiles to exsolve and form bubbles. The more viscous the magma, the harder it is for bubbles to escape, so expansion of bubbles in viscous magma can create a particularly violent explosion.
- Generally speaking, the volatile content of a magma goes up with increasing silica content, and so does viscosity. This means that felsic magmas are usually erupted explosively.
- Airfall deposits are produced by explosive eruptions. The thickness of the deposit and the size of the clasts both decrease away from the volcano. The pattern depends on the wind direction.
- Pyroclastic flows can be originated by collapse of an eruption column, by eruption directly from a vent, or by an avalanche from an unstable lava dome. They flow downhill, but their momentum can carry them up opposite-facing slopes.
- Calderas are formed by subsidence of the roof of a magma chamber when it is partly emptied during a large-volume (usually explosive) eruption, or a series of smaller eruptions.
- The shape of a volcano depends on the dominant kind of magma erupted. Mafic eruptions produce shield volcanoes (dominantly basaltic lavas), intermediate eruptions produce stratocones (interlayered andesitic lava and ash), felsic eruptions are commonly caldera-forming.
- A volcano will grow in an elongate fashion if magma is fed to the surface by fissures (volcanic rifts). Volcanoes are weak edifices, and may collapse.

6.6 Objectives for Section 6

Now you have completed this Section, you should be able to:

6.1 Describe and recognize the textural forms taken by flowing lava on dry land and under water, and the factors controlling these.

6.2 Describe some of the factors controlling the shape of lava flow fields.

6.3 Explain the links between magma type, gas content and the style of eruptions, and describe and recognize the main processes occurring during eruption and examples of the resulting deposits.

6.4 Recognize examples of different kinds of volcanoes and volcanic features, and explain how their shapes are controlled.

Now try the following questions to test your understanding of Section 6.

Question 6.6 Indicate the eruption products you would expect for mafic and felsic magma of high and low volatile content as indicated below by matching each case with one of (a)–(d) from the list:

(a) viscous domes; (c) fluid lavas;

(b) fire fountaining; (d) pyroclastic flows and airfall.

	Low volatile content	High volatile content
(a) mafic magma		
(b) felsic magma		

Question 6.7 Briefly describe three different sequences of events that could change the gross shape of a symmetric stratocone volcano. (*About 100 words*)

Question 6.8 Discuss how, on the basis of a single outcrop in the field, you could distinguish a deposit formed by a block-and-ash flow from an ignimbrite. (*About 100 words*)

Question 6.9 Complete the following Table, which is an aid to distinguishing sills from lava flows, by inserting ticks to indicate features you might see in cross-section in the field when studying an ancient volcanic area. Insert crosses to indicate features you would definitely not expect to find.

	Sill	Lava flow
chilled margin at top		
chilled margin at bottom		
columnar joints		
concordant base with local discordance		
concordant top with local discordance		
rubbly top		
rubbly bottom		
baking of overlying rock immediately above the contact		
baking of underlying rock immediately below the contact		

7 Igneous provinces

Most (but not all) volcanic and intrusive activity is a consequence of plate tectonics, so we will conclude our discussion of igneous processes with a brief look at this relationship. By the end, you should begin to understand why different volcanic processes are characteristic of different environments.

7.1 Constructive plate boundaries

Oceanic crust is created by igneous processes at constructive (i.e. divergent) plate boundaries. Burial of new ocean floor by marine sediments is so slow that close to the plate boundary there are vast bare expanses of basalt, mostly in the form of pillow lava and sheet flows. The ocean floor lacks the processes of deformation and erosion that are so useful for exposing the interior of rock bodies and structure in general to the geologist on land, so our knowledge of processes beneath the ocean floor relies on seismic studies, a limited amount of drilling, and sampling at anomalous sites such as transform faults and fracture zones where deeper levels are exposed at fault scarps. Fortunately, a slice of old oceanic crust and upper mantle tens or even hundreds of km in size occasionally escapes subduction and is thrust onto the continental crust where it can be studied by conventional geological techniques. An exotic slab of ocean floor of this kind is described as an **ophiolite** (sometimes called an ophiolite complex).

The picture that emerges from these lines of study is that the lowest part of the oceanic crust (which is much thinner than continental crust, about 8 km thick on average) is of plutonic rock equivalent in composition to the basaltic lavas forming the ocean floor.

❑ What rock name would you give to this rock type?

■ According to Block 2 Plate 6.12, the coarse-grained (i.e. plutonic) equivalent of basalt is gabbro.

The gabbro layer is believed to be constructed by episodic intrusion at the boundary between diverging plates. The magma solidifies slowly (hence its large grain size) and thereby adds to the edge of the plate. The ocean floor lavas are formed from magma that spills out when more magma is injected than can be accommodated in a subsurface magma chamber. The route taken by this magma to the surface is usually a vertical fissure, and each episode of lava extrusion is therefore recorded by the injection of a dyke. It is not uncommon for there to be a layer about 1 km thick between the lavas and the gabbros consisting of nothing but dykes, where each successive dyke has been intruded up the middle of a previous dyke or up the contact between two previous dykes. This layer is referred to as **sheeted dykes** or sometimes a sheeted dyke complex. As would be expected, the grain size of the dykes is medium (except for their chilled margins), and their composition is identical to the lavas. Their rock type is therefore described as dolerite (Block 2 Plate 6.12). The three-layer structure of oceanic crust is summarized in Figure 7.1.

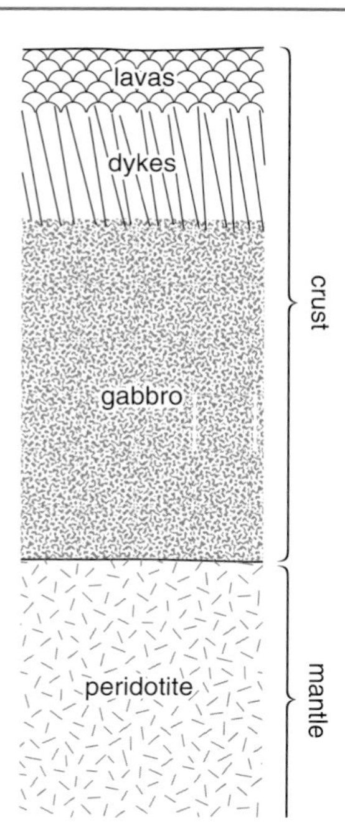

Figure 7.1 The three-fold division of oceanic crust into basalt lavas, dolerite dykes and gabbro.

An integral part of igneous processes at constructive plate boundaries is that seawater is drawn down into the hot young crust, where it is heated and rises back towards the sea-floor in a hydrothermal circulation system. The hot water reacts with some of the minerals in the crust and changes their chemistry (in particular, much of the pyroxene is replaced by the hydrous mineral amphibole). This is similar to the metasomatism described for a granite pluton in Section 5.1.

Given that there is all this magma at a constructive plate boundary, you should not be surprised that its composition is mafic. After all, mafic magma is what you would expect to find as a consequence of partial melting of the ultramafic mantle, and the mantle is the only potential melt source in such a setting. It is also natural that magma should escape upwards, because mafic magma is less dense than ultramafic rock. However, *why* the mantle should partially melt is not quite so obvious. Many students naïvely assume that there must be a linear heat source below the plate boundary. This is quite wrong, but by now you should know enough to deduce the truth.

Look back at Figure 2.8. This shows two plates diverging (pulled apart by forces acting at subduction zones, as remarked in Section 2.3.1) with asthenospheric mantle being drawn upwards from beneath the plate boundary at a rate sufficient to plug the gap.

❑ What will happen to the pressure experienced by this asthenospheric mantle as it is drawn upwards?

■ The pressure must drop as the depth decreases.

To see the implications of this, look at the *P–T* phase diagram for material of ultramafic composition (Figure 7.2). The mantle is actually very dry beneath constructive plate boundaries, so it is the *anhydrous* solidus that is relevant in controlling partial melting here. The diagram shows a likely path followed as asthenosphere is drawn upwards beneath a constructive plate boundary.

❑ According to Figure 7.2, what will happen to upwelling anhydrous mantle when it reaches a depth of about 90 km?

■ This is the depth at which it crosses the anhydrous solidus, so it will begin to melt.

The exact depth at which melting begins below a particular constructive plate boundary depends on factors such as the starting temperature, the speed of upwelling and the rate of cooling during uprise. The important point is that although upwelling mantle may cool slightly as it rises, the drop in pressure is

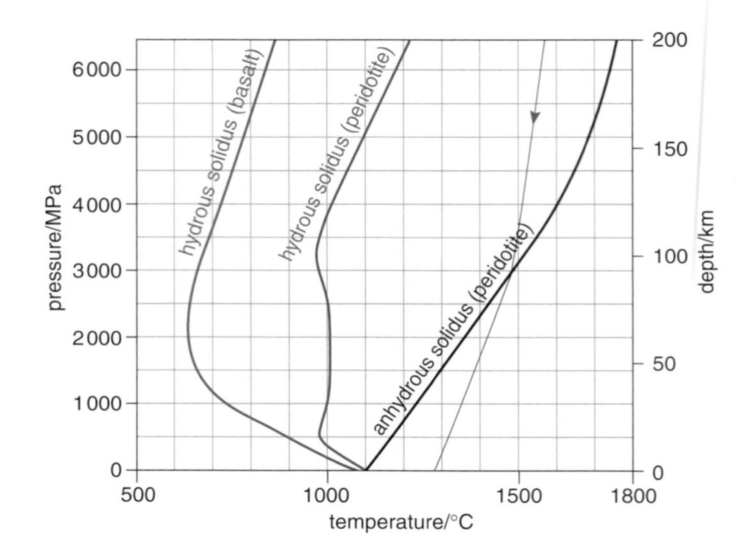

Figure 7.2 The path in *P–T* space (arrowed line) followed by mantle upwelling beneath a constructive plate boundary, relative to the anhydrous solidus for peridotite (i.e. ultramafic rock). Lines labelled 'hydrous solidus' show the solidus for basalt and for peridotite in the presence of 0.4% water. (*Note:* being 'hydrous' rather than 'water-saturated', the hydrous solidus for basalt does not coincide exactly with the water-saturated solidus in Figure 4.2.)

much more important, so that it can rarely avoid crossing the solidus into the partially molten field. Geochemical evidence indicates that this usually occurs at a depth of around 70–110 km, from whence the magma escapes upwards to form the crust. Mafic magma is therefore generated by partial melting of the ultramafic mantle entirely as a result of decompression melting, a process you met in Question 4.2.

7.2 Hot spots

A similar explanation accounts for the dominantly basaltic activity found at isolated **hot spots** around the globe. These are places where, independently of plate tectonics, a diffuse cylindrical upwelling, known as a **mantle plume**, rises from great depth within the mantle (perhaps as deep as the core–mantle boundary) and impinges on the base of the lithosphere. Although a certain amount of heat is brought up by such a plume, the main way of generating partial melts is decompression melting in the topmost part of the plume. Long-lived hot spots above mantle plumes are known both within oceans (where the magma is dominantly mafic) and within continents (where assimilation or partial melting of continental crust may give rise to more silica-rich magmas). Volcanoes above hot spots are often described as **intraplate volcanoes** (or within-plate volcanoes), because they are generally found a long way from a plate boundary.

The best-known example of an oceanic hot spot is Hawaii. Big Island's active intraplate volcanoes Mauna Loa and Kilauea (Figure 6.25) are sited directly above the plume. Their extinct fellows on the same island, the other islands of the Hawaiian archipelago, and a chain of seamounts (subsea volcanoes) that can be traced for 3400 km to the north-west represent increasingly old volcanoes that have been carried away from the hot spot by the north-westward motion of the Pacific Plate over the effectively stationary plume (Figure 7.3). Continental intraplate igneous activity can be found in the southern Sahara, where half a dozen mainly stratocone volcanoes make up the Tibesti mountains, rising to 3400 m. There is no linear chain of extinct volcanoes here because the African Plate is hardly moving relative to the underlying mantle plume.

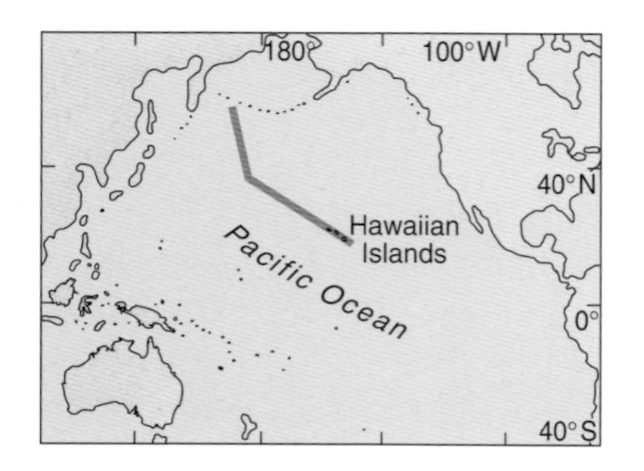

Figure 7.3 The location of the Hawaiian Islands and the chain of seamounts (coloured line) extending north-westwards from the hot spot.

7.3 Flood basalts

Intraplate volcanism fed by decompression melting has occurred on a much grander scale many times in the past. Usually, but not always, this is a prelude to continental rifting and occurs at stage (b) in the sequence indicated in Figure 2.9. Normally the amount of magma produced is slight, but where the incipient rift happens to overlie a mantle plume a much greater volume of basaltic partial melt is generated. This escapes through dykes feeding a series of lava flows, individually tens of metres thick but adding up to a cumulative thickness of around a kilometre. These flows are described as continental **flood basalts**.

The world's most famous example is perhaps the Deccan Traps basalts that cover a large part of western India (Figure 7.4). A volume of magma of more than one million km^3 was erupted in less than half a million years, 65 million years ago when India was beginning to break away from Madagascar.

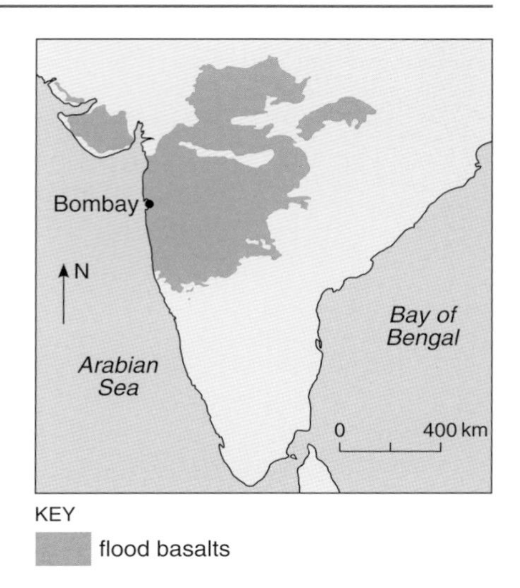

Figure 7.4 Map showing the present-day extent of the Deccan Traps continental flood basalts in India. The average total thickness is about 1 km. They cover an area of half a million km^2 today, but prior to erosion, and taking into account what is below sea-level, their original extent may have been 1.5 million km^2.

Continental flood basalts also mark the rifting of the British Isles from Greenland about 57 million years ago in the early Tertiary (heralding the opening of the North Atlantic). Isostatic subsidence of the stretched and thinned crust underlying these flood basalts means that most of the area is now below sea-level but notable examples on the fringes can be seen in western Scotland and Northern Ireland (Block 2 Figure 6.1 and Plate 6.1, and examples in Video Bands 4 and 5).

Some flood basalt provinces are known on oceanic crust too, the largest being the Ontong–Java Plateau in the western Pacific. This cannot be a consequence of continental rifting, but clearly represents a prodigious volume of basaltic melt generated in an oceanic intraplate setting. The cause is enigmatic, but is presumably connected with a mantle plume.

Flood basalts seem to be erupted at random intervals at random places on the globe, and are among the clearest demonstrations that our planet is not in a uniform steady state but undergoes episodes of enhanced activity. The lava erupted in a large flood basalt province would clearly have devastating local effects, but the amount of volcanic gas liberated over a comparatively short period could also have significant effects on global climate.

7.4 Destructive plate boundaries

Fortunately for us, there are no indications that eruption of another suite of flood basalts is imminent. Most of the world's on-land volcanoes occur above destructive plate boundaries on the edges of continents and in oceanic island arcs where andesitic stratocones of the kind described in Section 6.4 predominate. In this Section, we will examine how subduction leads to volcanism. Once again, phase diagrams hold the key.

Look back at Figure 2.6. This shows oceanic lithosphere being subducted, and volcanoes situated above the subduction zone. Clearly, melt must be being generated somewhere below the volcanoes. Consider what happens to the downgoing plate, usually referred to as the **slab**, during subduction.

❑ What will happen to the pressure and temperature in the slab?

■ The pressure will increase according to depth. As it gets deeper, the slab comes into contact with progressively deeper and warmer parts of the mantle. Heat will be conducted into the slab, which will cause it to warm up.

Pressure is transmitted instantaneously, so the pressure in the slab must always be the equilibrium pressure for the depth, but the rate at which heat can be conducted into the slab is limited by its thermal conductivity. Therefore, the slab will warm up slowly, as indicated in Figure 7.5.

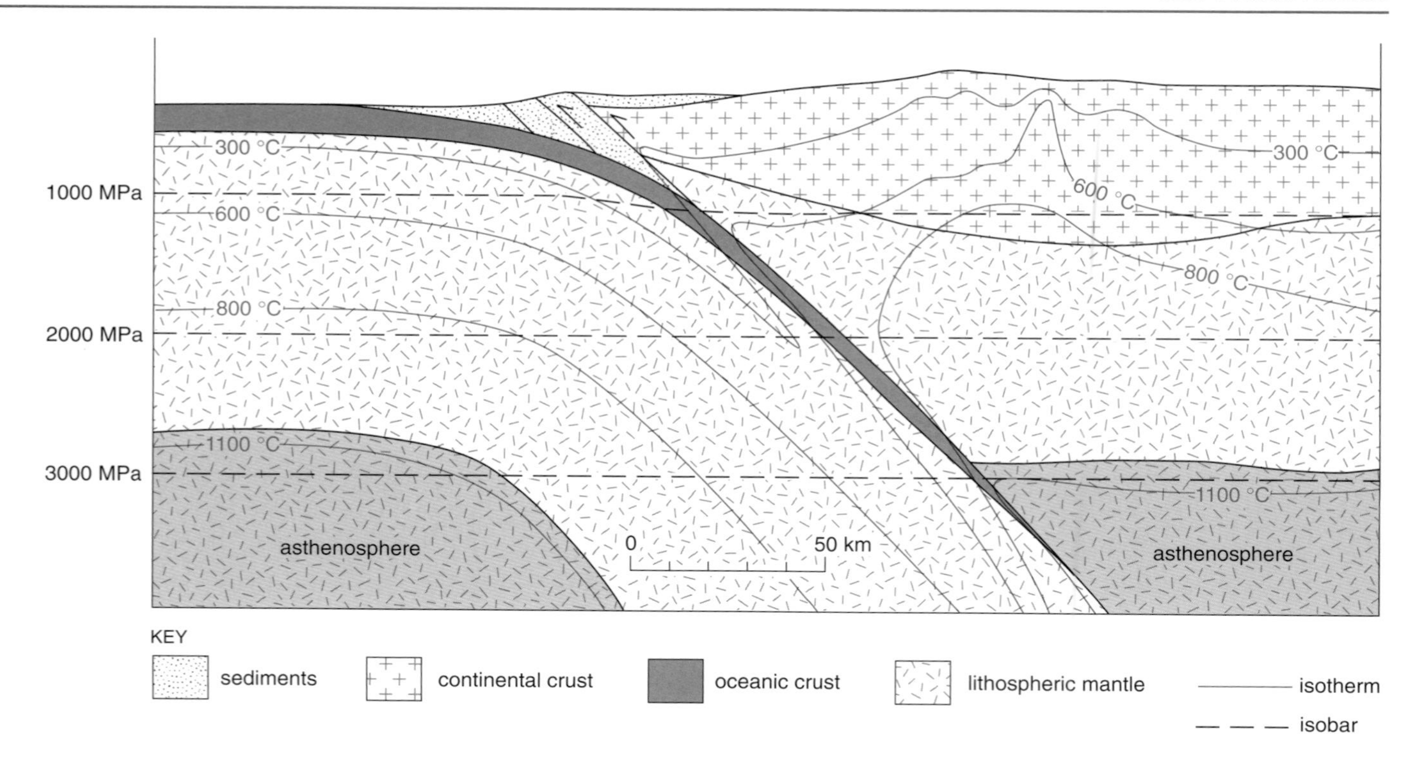

Figure 7.5 Cross-section through a destructive plate boundary, showing predicted temperature contours (isotherms) and lines of equal pressure (isobars). Downward deflection of isotherms is caused by the subduction of the cold slab. Upward deflection of shallow isotherms above the subduction zone is because of heat carried by rising magma. Details depend on the rate of subduction, but the pattern is representative.

Although the very uppermost part of the oceanic crust could be made to melt because of heat conducted into it from the hot mantle it encounters, we should look to the pressure change as the main cause of partial melting. If you look at Figure 4.2 you will see that the anhydrous solidus for mafic rock slopes the wrong way, so that an increase in pressure would drive conditions further into the 'solid only' field. However, although oceanic crust is created by partial melting from an anhydrous source (as you saw in Section 7.1), by the time it is subducted anhydrous conditions no longer apply.

- ❑ What reasons can you think of?
- ■ The most obvious is that a subduction zone begins under water, so the upper part of the slab must be wet. Seawater will permeate cracks and fissures in the oceanic crust, and the veneer of sediments overlying the lavas (which may be subducted with the slab) will be wet too. Also, as you saw in Section 7.1, much of the original pyroxene in the crust of the slab will have been hydrothermally replaced by amphibole, which is a hydrous mineral.

The combination of trapped water and the hydrous minerals in the hydrothermally altered crust means that the *hydrous* solidus is the relevant phase boundary for partial melting in the slab. On Figure 7.2, you can see that the hydrous solidus for basalt slopes in the appropriate direction so that an increase in pressure can lead to partial melting even without a particularly high temperature. To reassure yourself of this, try Question 7.1.

Question 7.1

(a) At 2000 MPa, what is the temperature in the crustal part of the slab according to Figure 7.5?

(b) What does Figure 7.2 tell you about the state of mafic rock under these conditions?

The conditions described in Question 7.1 are those at which significant volumes of partial melt begin to be produced. Partial melting will progress further as the slab continues its descent. Thus, intrusion and eruption of intermediate rock types above subduction zones can be explained as a direct consequence of partial melting of the wet mafic crust in the downgoing slab. The crust of the overriding plate has no direct role in the process. However, magmas of basaltic and felsic composition are also intruded and erupted at destructive plate boundaries, granites and felsic ignimbrites being particularly notable. This is because there are other melt-generating processes operating too, and because fractional crystallization and assimilation (Section 4.3) can cause the silica content of magma to increase during ascent. The full interplay of processes is complex, and is not completely understood, but the essential story appears to be as follows.

When the slab reaches a pressure of about 2500 MPa (equivalent to a depth of about 80–90 km), the minerals in the oceanic crust undergo a phase transition to produce a denser mafic rock type called **eclogite**, which consists of garnet and pyroxene. Both these minerals are anhydrous, so the water previously held within the amphibole and other hydrated minerals is driven out. In addition, pressure will squeeze out the water that was originally carried down in cracks and fissures. The only place for the water to go is up, and it must first pass through the overlying wedge-shaped region of mantle belonging to the overriding plate (known as the **mantle wedge**). Just as addition of water to mafic rock can lead to partial melting (Question 4.3), so does addition of water to the previously anhydrous ultramafic material in the mantle wedge, in this case generating mafic magma. Similarly, escaping water reaching the base of the crust can change local conditions from anhydrous to hydrous, and so partial melts of andesitic, dacitic or rhyolitic composition can arise. The suite of processes generating and changing the composition of magma is summarized in Figure 7.6.

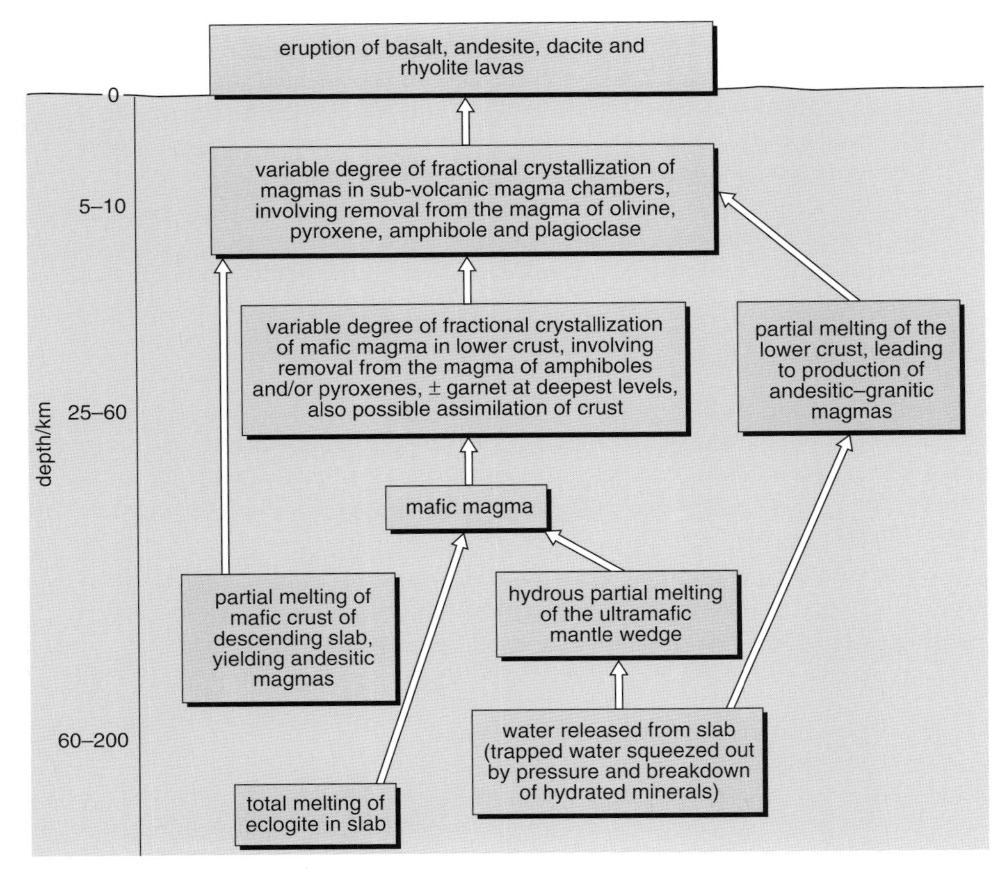

Figure 7.6 Possible processes involved in the generation and evolution of magmas at destructive plate boundaries.

7.5 Continental collision

There remains only continental collision to complete our survey of the main settings for igneous processes. Figure 2.7b showed that subduction stops after two continents have collided, at which time the magma-generating processes indicated in Figure 7.6 must terminate too. The rates of magma ascent described in Section 5.1 imply that emplacement of plutons and eruption of volcanoes should die out within a million years or so of this. However, it is found that plutons, especially granites, are emplaced in continental collision zones for several tens of millions of years after a collision. The Shap Granite that you studied in Activity 5.1 (and indeed the Lake District batholith in general) is an example of this type, having been intruded in the aftermath of a collision that united England and Scotland in the Silurian.

Geochemical arguments can be used to demonstrate that there is a substantial contribution from the mantle in some post-collision granite magmas, which is not very well understood. Other granite magmas can be explained more simply by considering what collision does to the crust. Compressional deformation of the leading edge of the overriding continental crust combined with partial subduction of the leading edge of the continental crust on the subducted plate means that the crust near the suture zone is much thicker than before collision (Section 2.3.1). For example, the continental crust below the Himalayas (the site of a suture zone between India and the main mass of Asia) is about 70–80 km thick, or about twice the average continental thickness.

Figure 2.7b does not show the complex interplay between thrusting and compressional folding in thickening the continental crust, but the details have little effect on magma generation.

❑ Can you remember the relative contributions of continental crust and mantle to radiogenic heating within the Earth?

■ Volume for volume, continental crust generates heat at a rate orders of magnitude greater than the mantle (see Section 2.4).

❑ Which of the following rock types do you think will melt first to form a granite magma because of the enhanced rate of heat generation in the aftermath of a continental collision: granite, quartz sandstone, limestone, mudstone (consisting of micas, clay minerals and quartz)?

■ We need a rock consisting of a hydrated assemblage of minerals containing the necessary elements. The mudstone and the granite both contain the right elements, but the high proportion of hydrated minerals in mudstone (clays and micas; Block 2 Sections 4.5.1 and 4.5.2) would cause it to begin to melt before a granite. Limestone and quartz sandstone do not contain the right ingredients to generate a granite magma.

The origin of many of the granites in collision zones is believed to be melting of muddy sediments originally from the upper crust or partial melting of other less silica-rich crustal rocks (average continental crust is intermediate in composition, see Question 2.2) triggered by the enhanced rate of heat generation in the thickened crust.

We take a closer look at how heat flow is affected by crustal thickening in Section 8.4, but for now perhaps you will be relieved to have at last found a melting mechanism that relies on heat rather than pressure changes!

7.6 SUMMARY OF SECTION 7

- At constructive plate boundaries and intraplate volcanoes, mafic magma is generated by decompression melting of anhydrous upwelling mantle.
- At destructive plate boundaries, partial melting occurs in hydrous conditions. The downgoing slab's oceanic crust partially melts to produce magma of intermediate composition, mainly because of the pressure increase as it descends. Water escaping from the slab hydrates the mantle wedge and the overlying crust to produce partial melts of mafic and intermediate-felsic composition, respectively. Melts may become more felsic during ascent, mainly as a result of fractional crystallization but perhaps also through crustal assimilation.
- The rate of radiogenic heat production in the thickened continental crust resulting from a continent–continent collision causes sufficiently high temperatures for magmas of mainly felsic composition to be produced by partial melting within the crust.

7.7 OBJECTIVES FOR SECTION 7

Now you have completed this Section, you should be able to:

7.1 Describe the most common types of igneous activity at constructive and destructive plate boundaries and in intraplate settings, and explain how the magma is generated and how its composition may evolve during ascent.

Now try the following questions to test your understanding of Section 7.

Question 7.2 Describe two ways in which magma ultimately erupted as a felsic ignimbrite could be generated near a destructive plate boundary. (*About 150 words*)

Question 7.3 According to the conditions portrayed in Figure 7.5, and with reference to the phase boundaries on Figure 7.2, explain whether or not you would expect partial melting to occur in the mantle immediately overlying the slab at a depth equivalent to a pressure of 3000 MPa. (*Two or three sentences*)

8 METAMORPHISM

8.1 INTRODUCTION

As described in Section 2.4, the Earth is heated from within by radiogenic heat from decay of the heat-producing isotopes of K, U and Th. This is why temperatures increase with depth inside the Earth. The formation of granites by melting crustal rocks is now familiar to you from the previous Section, but heating the Earth's crust does not always result in melting. As rocks become warmer, their constituent minerals recrystallize and form different minerals that are stable at higher temperatures. Deeper in the Earth's crust (or mantle), the mass of the overlying rocks results in higher pressures imposed on the rocks and new minerals form that are stable at these pressures. Moreover, the shapes and alignments of minerals change in response to increasing pressures as rocks reorganize themselves to take up smaller volumes or different shapes. The changes in the mineralogy and in the shape of minerals that make up a rock when subjected to changes in temperature and/or pressure are known as metamorphism.

Broadly, the study of metamorphic rocks comprises three main aspects, which should be considered together wherever possible:

1 *Evaluating the conditions of temperature and pressure to which the rocks were subjected during metamorphism.* We have direct evidence that new minerals form at given temperatures and pressures from many laboratory experiments performed over the past 40 years. Rocks of different compositions are heated and squeezed in a laboratory under specified pressures and temperatures, and the resulting minerals are compared with those observed in naturally occurring metamorphic rocks. The conditions of pressure and temperature that produced those rocks are assumed to be the same as those imposed during the experiment. The assumption that rocks in the laboratory will behave similarly to rocks in the Earth's crust or mantle may not always be valid, though. Natural rocks are rather complex chemical systems. In addition to the major elements that make up the more abundant minerals, they contain a large number of minor elements that, although present in minute concentrations, can have a strong effect on the stability of metamorphic minerals. Equally importantly, minerals in natural rocks normally form in the presence of a fluid phase (usually a mixture of H_2O and CO_2), and the composition of this fluid influences which minerals are stable under the particular P–T conditions. Unfortunately, only rarely can we sample the fluid phase that was present during the metamorphism of any rocks in which we may be interested. The best we can usually do is to infer compositions from the minerals we observe. Finally, there is the problem of time. Natural rocks usually have millions of years in which to react to new metamorphic conditions; but in the laboratory, the available time-scale is restricted to days or perhaps weeks. This is most important at low temperatures, because under such conditions the rates of reaction are very slow, and it is difficult to achieve equilibrium in the laboratory experiments. At higher temperatures, reactions proceed more rapidly, and so equilibrium can be achieved in a comparatively short space of time.

2 *Thermal evolution and its relationship with the tectonic environment.* In trying to interpret the significance of the P–T information that we learn from studying metamorphic minerals, we must make sure it fits in with all the other geological evidence available. Heat for metamorphism comes from both internal sources (largely radiogenic heat), and from external sources such as the intrusion of hot magma. The precise combination of these sources determines the way that temperature increases with depth. This is recorded by the distribution of metamorphic minerals, and reflects the tectonic setting in which the rocks have formed. When metamorphic studies are combined with structural studies, the tectonic setting can often be identified, not only during the highest metamorphic temperatures that the rocks have experienced but also whilst the rocks were cooling during their ascent towards the surface during uplift and subsequent erosion. This process is called **exhumation**.

3 *The relationship between metamorphism, as seen in the growth of new minerals, and deformation events recorded in large-scale and small-scale structures.* A careful study of the mineral textures seen in thin sections and hand specimens provides the bridge between an estimate of metamorphic conditions and an understanding of the geological events that caused them: without it, we should learn little of the history of a metamorphic terrain. Such studies are based on textural evidence of the sort discussed in Section 11.5.

In practice, the subject of metamorphism is limited to those transformations that take place between the near-surface zones of sedimentation and cementation of sedimentary rocks, and zones deep in the Earth's crust and mantle where partial melting begins. Such boundaries are gradational, and so inevitably there are areas of overlap in the study of the mineralogical and chemical changes that occur during the compaction and lithification of sediment, and the study of igneous petrology, where rocks start to melt. These two examples are of very low **metamorphic grade** and very high metamorphic grade respectively.

The simplest definition of metamorphic grade is that higher grades of metamorphism reflect conditions of higher temperatures. It is a rather loose term but will be defined more precisely later.

Activity 8.1

Before considering the nature of metamorphic reactions in more detail, it may be useful to re-examine the metamorphic rocks from the Home Kit, as described in this Activity.

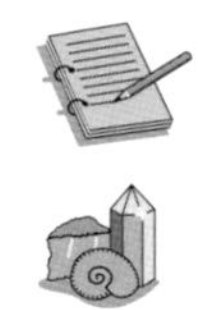

8.2 METAMORPHIC REACTIONS

Rocks are clearly solid objects and do not appear to react, at least at room temperature and atmospheric pressure. However, minerals, like most other chemical compounds, will react if temperatures and pressures are changed sufficiently. In fact, all mineral assemblages in metamorphic rocks result from chemical reactions that take place as the rock undergoes metamorphism. These can be expressed either in terms of mineral names or chemical formulae.

For example, consider the reaction:

$$\text{muscovite(s)} + \text{quartz(s)} \longrightarrow \text{alkali feldspar(s)} + \text{sillimanite(s)} + \text{H}_2\text{O(g)} \quad (8.1)$$

H_2O will be a gas or a liquid depending on the conditions of the reaction. The minerals muscovite and quartz are the **reactants** and alkali feldspar and sillimanite are the mineral **products**. The same reaction can be expressed by an equation using the chemical formulae of the four minerals, and the fluid phase.

$$\underset{\text{muscovite}}{KAl_3Si_3O_{10}(OH)_2} + \underset{\text{quartz}}{SiO_2} \longrightarrow \underset{\text{alkali feldspar}}{KAlSi_3O_8} + \underset{\text{sillimanite}}{Al_2SiO_5} + H_2O \quad (8.2)$$

There are two properties of chemical reactions we now need to introduce in order to understand metamorphic reactions better. First, any chemical reaction involves a change in **entropy** (represented by the symbols ΔS, pronounced 'delta S') between reactants and products. Entropy reflects the ordering of the atomic structure of matter. A system with zero entropy is perfectly ordered whereas the same system with high entropy is more disordered. In Block 2 Section 2.1, we established that a liquid or a gas is more disordered than a crystalline solid at a given pressure (Block 2 Figure 2.1). Now consider the chemical reaction involved in dissolving sugar in a cup of coffee. The reactants are sugar (a crystalline solid) and coffee (a liquid); the product is entirely liquid (sweet coffee), so the system has increased its atomic disorder and therefore increased its entropy.

❑ To take another example, which has the lowest entropy: ice, water or steam?

■ Clearly ice, with a crystal structure, is the most highly ordered and so has the lowest entropy.

Secondly, any chemical reaction involves a change in **molar volume** as represented by the symbols ΔV ('delta V'). This is defined as the change in volume between the sum of the reactants and the sum of the products. A solid, liquid or gas with a small molar volume has a closely packed atomic structure and is dense; conversely, one with a large molar volume is less dense. So a gas has a larger molar volume than a liquid which in turn has a larger molar volume than a solid. When graphite is transformed into diamond at high pressures, there is a net increase in density and so a decrease in the molar volume.

❑ Returning to reaction 8.1, is the value of ΔS positive or negative? (You can assume that the four minerals involved have similar values of entropy.)

■ H_2O is the only gas involved, and gas has a less ordered structure than any solid or liquid and so has the highest entropy of the three physical states – solid, liquid and gas. Consequently, entropy of the products is greater than the entropy of the reactants. In other words, entropy increases as the reaction proceeds to the right and so ΔS is positive.

❑ Is the value of ΔV positive or negative in reaction 8.1?

■ It is positive as two solids have reacted to form two other solids and a gas, which is much less dense (and so has a large molar volume) than any solid.

Entropy may also be used to derive a more satisfactory definition of metamorphic grade. Earlier, we stated that 'higher' grade is usually taken to mean higher temperature. We may now define metamorphic grade in terms of the entropy change (ΔS) of the metamorphic reactions concerned. *Increasing grade involves an increase in entropy of the metamorphic system* (including of course any gas phase).

❑ In Equation 8.1, which assemblage will be present at higher metamorphic grade: (i) muscovite and quartz or (ii) sillimanite, alkali feldspar and H_2O?

■ Assemblage (ii), sillimanite, alkali feldspar and H_2O: this assemblage is less ordered, so has the higher entropy.

Most metamorphic reactions take place as the metamorphic grade increases, and are called **prograde reactions**; they tend to decrease the degree of ordering in the mineral system concerned.

Question 8.1 Another common metamorphic reaction is that of

$$\text{tremolite(s)} + \text{calcite(s)} + \text{quartz(s)} \longrightarrow \text{diopside (s)} + H_2O(g) + CO_2(g) \qquad (8.3)$$

where tremolite and diopside are an amphibole and a pyroxene respectively.

Determine (a) which assemblage has the lower entropy, and (b) which assemblage would be stable at a higher grade of metamorphism.

For the muscovite+quartz reaction (Equations 8.1, 8.2) at a pressure of 150 MPa, temperatures of more than 550 °C are needed for muscovite and quartz to react to form alkali feldspar, sillimanite and H_2O. If alkali feldspar, sillimanite and H_2O are taken below 550 °C again, will they react together and revert to muscovite and quartz? If so, how can it be that alkali feldspar and sillimanite are often found in the same rock at room temperatures?

There are two main reasons:

1 One of the products of the prograde reaction is a gas, H_2O, and the short answer is that it is not likely to stay around and wait for the rock to cool. Thus, if alkali feldspar and sillimanite are unaccompanied by H_2O at temperatures below 550 °C, they obviously cannot revert to muscovite and quartz.

2 The rates at which reactions take place vary enormously. In particular, they increase exponentially with increasing temperature, so that many reactions are extremely slow at low temperatures. Thus, if an assemblage is cooled rapidly, it may not have enough time to react and form low-temperature minerals.

In general, if conditions are such that reactions do take place during cooling, they are called **retrograde reactions**; these occur as rocks attempt to readjust to the lower grades of metamorphism. The best textural evidence for retrograde metamorphism is to find a relatively high-grade mineral surrounded, and partially replaced, by a mineral that is stable at lower metamorphic grades. For example, Plate 8.1 shows garnet being replaced by chlorite. Moreover, since both rarely coexist in equilibrium, we may conclude that the photograph preserves a record of **textural disequilibrium** between the garnet and the chlorite because the reaction has not gone to completion. Retrograde reactions tend to occur during slow cooling in rocks that contain a H_2O-rich fluid phase. The reason that some garnet is preserved in Plate 8.1 is either that the reaction ran out of fluid, or the rock cooled too fast for the reaction to go to completion.

Unfortunately, because the experimental problems are acute, very little is known about the actual rates of specific metamorphic reactions between coexisting solids. An exception is the change of aragonite to calcite. These two minerals are polymorphs: both have the same composition ($CaCO_3$) but different crystal structures, and the change from one to the other takes place in about a year at temperatures above 400 °C. The reaction has been studied experimentally at these temperatures and results extrapolated to lower temperatures by plotting a graph of temperature against the time taken for all the aragonite to change to calcite (Figure 8.1). Note that such extrapolations can introduce comparatively large errors.

❑ Why is the experiment not conducted at lower temperatures?

■ If you look at the time axis, you can see that it would take approximately a *thousand* years at 325 °C for all the aragonite to change to calcite, a *million* years at 270 °C, and a *thousand million* (or one billion) years at 240 °C.

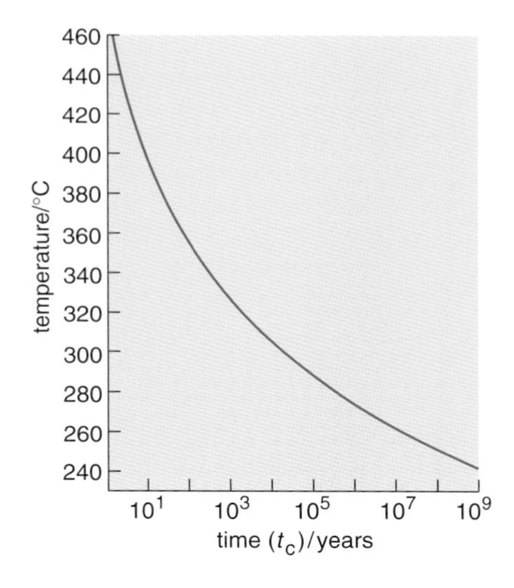

Figure 8.1 Time (t_c) for complete change from aragonite to calcite as a function of temperature.

Those are long tracts of time, even for a geologist! However, there is one simple way of speeding up the change – just add water. The experiments used to construct Figure 8.1 were done in the absence of water, and if water is present the times are greatly reduced. Even at 50 °C, all the aragonite is changed to calcite in a mere 10^6 years, a consequence of which is that aragonite is very rarely found in fossils or as cement in ancient rocks.

Why should water speed up the change bearing in mind that it cannot contribute chemically in a reaction in which one polymorph of $CaCO_3$ goes to a second polymorph of $CaCO_3$?

Water is an excellent **catalyst** and is particularly important in speeding up metamorphic reactions that otherwise might not take place in the time available. In completely dry rocks, chemical changes can only occur by the slow process of ionic diffusion (the passage of ions through a solid crystal lattice). When water is present, it tends to spread itself along grain boundaries as an intergranular film, providing a network of chemical 'arteries' in which ions may move rapidly in solution and so speed up the metamorphic reactions.

This discussion of the muscovite+quartz reaction and the structural change from aragonite to calcite highlights three important general points:

1 Many prograde metamorphic reactions release H_2O and/or CO_2. Because these products are mobile and tend to move away from the rocks in which they were generated, they are often unavailable for retrograde reactions that might otherwise take place as the temperature falls.

2 Rates of reactions tend to be slow, but they increase exponentially with increasing temperature and are also speeded up by the presence of water.

3 Conversely, since many prograde metamorphic reactions release H_2O and CO_2, both tend to be driven off as metamorphism continues. Thus, by the time the highest grades are reached, almost all the fluid will have left the system, and recrystallization and the growth of new minerals will have sealed off all the intergranular spaces, making the rock almost impermeable. Not only are H_2O and CO_2 not present to take part in reverse reactions, they are also unavailable as catalysts. Thus, such 'dry' rocks will then survive cooling more or less unaltered, because the rates of reaction are limited by the very slow rates of ionic diffusion. So high-grade mineral assemblages become 'frozen' in and preserved at low temperatures.

High temperature minerals can be present at low temperatures but do not lie within their stability field when plotted on a phase diagram. Minerals that exist outside their stability fields are described as being metastable (Block 2 Section 4.6), which means that while they are not in a state of change, they can be made to change by some sort of impetus. For metamorphic assemblages, the impetus either comes from further metamorphism, when the rock is reheated and/or squeezed under a new set of *P*–*T* conditions, or from weathering, when the rock is attacked by atmospheric agencies and its minerals are transformed into those that are in equilibrium at the temperatures and pressures present at the surface of the Earth – like clay minerals in sediments.

❑ How do we know that metamorphic rocks were in fact formed at high temperatures and pressures?

■ From experiments designed to reproduce naturally occurring mineral assemblages under conditions of known pressure and temperature.

The results of such experiments are most simply illustrated using a phase diagram. In the experiments on which phase diagrams are based, there are three things that can be varied easily – the chemical composition of the sample, the temperature, and the pressure. The minerals that form will depend on all three. It is obviously difficult to illustrate how they all vary on a two-dimensional diagram, and for convenience we usually keep one constant and show what happens as the other two vary. In most metamorphic experiments, it makes sense to work with a fixed composition, and to plot the results on a graph of pressure against temperature. Always remember, however, that such a *P*–*T* phase diagram only illustrates results for a particular composition; if you alter the composition, the details of the phase diagram will change.

The phase diagram for the composition Al_2SiO_5 illustrates the different pressures and temperatures at which the three polymorphs (kyanite, sillimanite and andalusite) exist under equilibrium conditions (Figure 8.2). Several points should be emphasized:

1 A *phase* is formally defined as something that differs chemically and/or physically from the rest of the system being considered. In this case, kyanite, sillimanite and andalusite are three different phases, because although they share the same chemical composition they have different crystal structures (i.e. they are polymorphs).

2 At equilibrium, the minerals kyanite, sillimanite and andalusite each exist over a range of different pressures and temperatures (Figure 8.2). These are the conditions under which each mineral is stable, and the area depicting these conditions on a phase diagram is called its stability field. (Note also that to describe a mineral as 'stable' is synonymous with saying it is 'at equilibrium'.)

3 If a phase is present under conditions outside its stability field, we refer to it as being metastable.

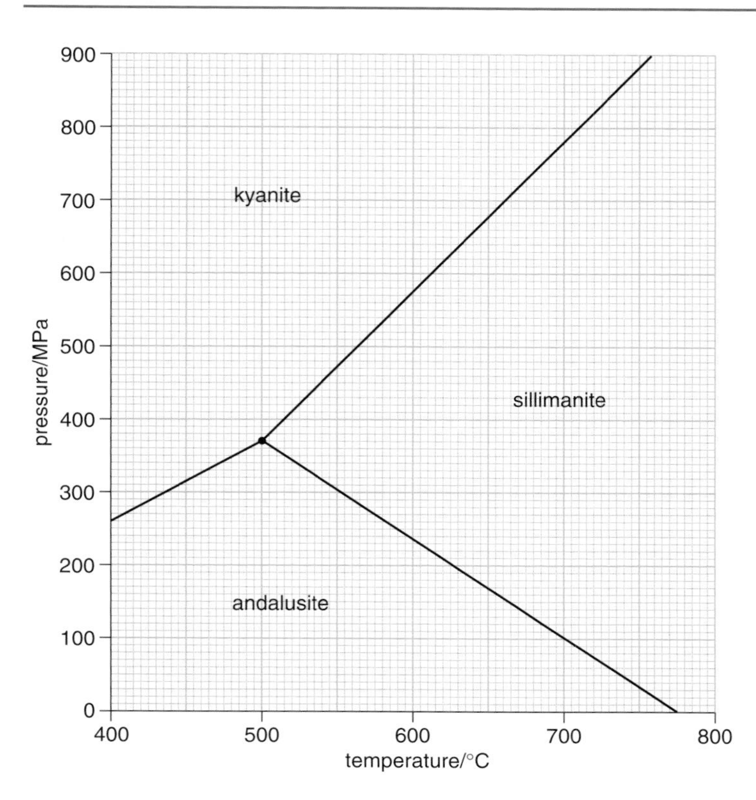

Figure 8.2 Phase diagram illustrating the stability fields of three aluminium silicates: andalusite, kyanite, and sillimanite.

4 The lines that mark the boundaries between the stability fields of kyanite, sillimanite and andalusite are phase boundaries. These lines mark the only places on this phase diagram where two of these minerals can exist together *at equilibrium*.

If either pressure or temperature is changed so that conditions move off the phase boundary, only one of the minerals becomes stable. We have seen how such changes can take a very long time to complete (Figure 8.1).

Question 8.2 Use the Al_2SiO_5 phase diagram (Figure 8.2) to answer the following questions.

(a) Is andalusite stable at 300 MPa and 700 °C?

(b) Which phase is stable at 800 MPa and 500 °C?

(c) At what temperature do kyanite and sillimanite coexist in equilibrium at a pressure of 700 MPa?

(d) Kyanite and andalusite were reported together in the same sample at 400 °C at atmospheric pressure (0.1 MPa). Which phase is metastable (i.e. lies outside its stability field)?

Your answers to Question 8.2 demonstrate that in the case of the kyanite–andalusite–sillimanite system, it is comparatively easy to make very general statements about the conditions of pressure and temperature, depending on which mineral was present at equilibrium; however, it is difficult to be specific. We can do better if we find a sample in which two of the minerals occur together in equilibrium, since that indicates that conditions were somewhere along the line of the phase boundary. But can we improve on that? Can we say precisely where along the phase boundary the minerals from a sample were in equilibrium? The short answer is no – unless help is available from elsewhere. What we need is some other mineral with a different stability field to be present in the same rock.

❑ What other reaction have we looked at which also involves sillimanite?

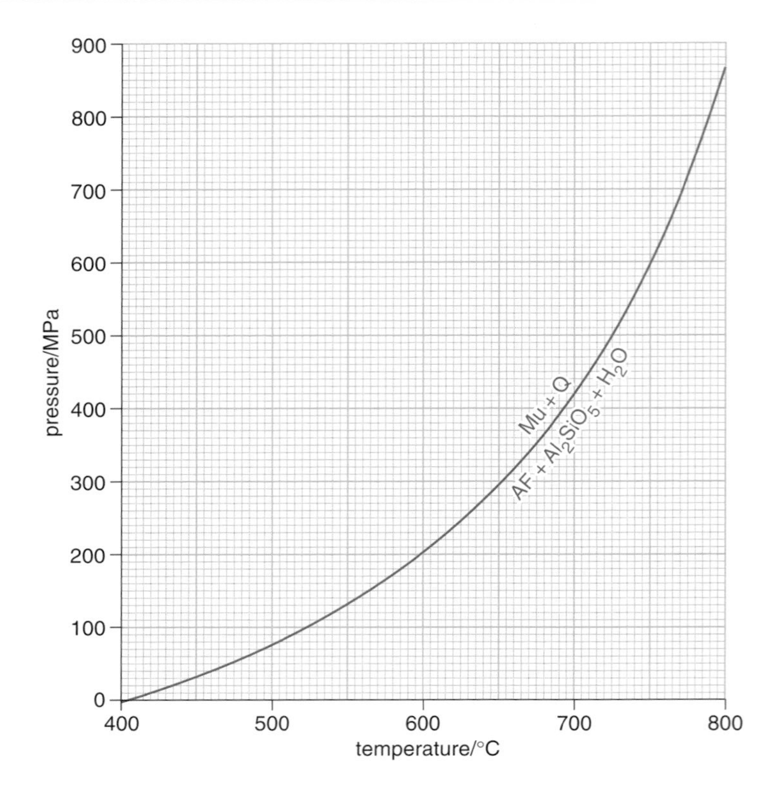

Figure 8.3 Phase diagram for the reaction muscovite (Mu) + quartz (Q) ⟶ alkali feldspar (AF) + Al_2SiO_5 + H_2O.

■ The one in Equations 8.1 and 8.2 in which muscovite and quartz react to form sillimanite, alkali feldspar and H_2O.

The phase diagram for this reaction is presented in Figure 8.3, and in this case, the phase boundary marks the boundary between the stability field of muscovite and quartz and that of Al_2SiO_5, alkali feldspar and H_2O. Note that we have replaced the term sillimanite with the formula Al_2SiO_5 since other polymorphs of this composition are stable within the *P*–*T* conditions shown. The boundary defines the conditions at which the five phases can coexist together in equilibrium. Because the gradient of this slope is fairly steep we can say that for normal crustal pressures ($P < 1000$ MPa), all five phases will coexist over a temperature range of about 400–800 °C. But how can we obtain a more precise estimate?

Figure 8.4 combines the phase diagrams for sillimanite–kyanite–andalusite and for muscovite–quartz–sillimanite–alkali feldspar–H_2O. By using all four phase boundaries, the diagram becomes subdivided into smaller areas of pressure and temperature (A, B, C, D and E) in which particular combinations of minerals are stable. We see, for example, that muscovite and quartz only coexist in equilibrium with andalusite in area A, rather than in the larger area (A + D + E) as implied by Figure 8.3.

Secondly, there is now only one point on the diagram where six phases can occur together. So, if muscovite, quartz, alkali feldspar, andalusite, sillimanite and H_2O are found in equilibrium in one sample, it can only have crystallized at a particular pressure and temperature.

❑ What would the pressure and temperature be?

■ These six phases only coexist at about 610 °C and 220 MPa (where the red and black phase boundaries intersect on Figure 8.4).

In summary, it should be apparent that:

1 It is not possible to obtain precise estimates of pressure and temperature on every sample of metamorphic rock. Usually, all you can say is that the rock crystallized somewhere in the stability fields of its particular minerals.

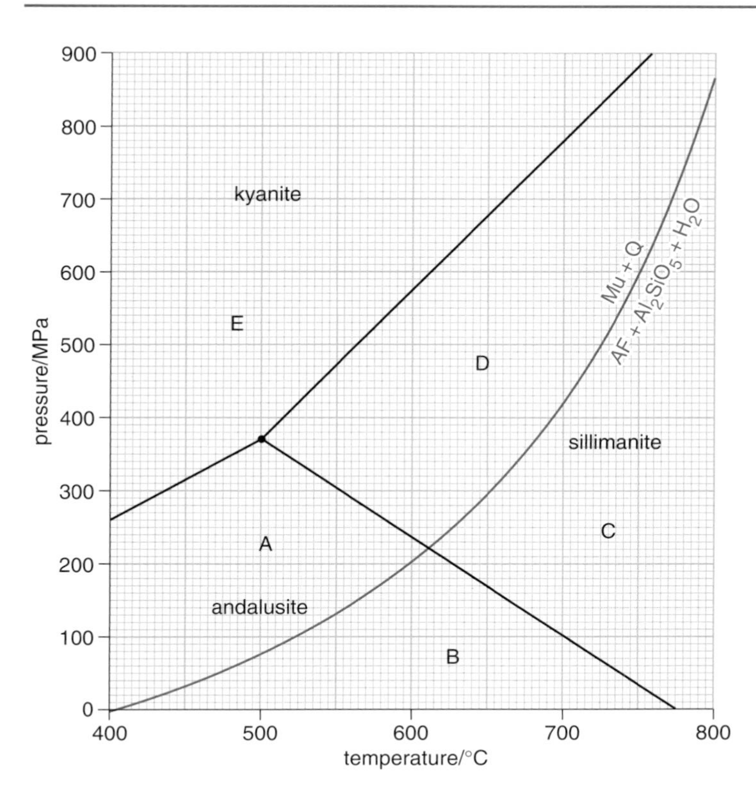

Figure 8.4 Phase diagram combining information from Figures 8.2 and 8.3; a simple petrogenetic grid.

2 To obtain a reasonable estimate of metamorphic conditions, it is necessary to seek out rock specimens that contain certain key metamorphic minerals and so may reflect conditions on a particular phase boundary. In some areas, suitable samples may be scarce or non-existent.

In general, it is necessary to combine information from different metamorphic reactions in an attempt to limit the possible range of pressure and temperature at which the rock could have crystallized. The grid-like pattern that is formed as more and more boundaries are added to the phase diagram is called the **petrogenetic grid**, and Figure 8.4 is a simple example.

There is another way of determining pressure and temperature of which you should be aware. In contrast to the minerals we have discussed so far in this Section, many minerals have a range of compositions; the plagioclase feldspars for example range in composition between albite and anorthite by the interchange of NaSi and CaAl (Block 2 Section 4.6.2). Reactions involving such minerals will not occur along a line on a phase diagram but be spread out across a range of P–T conditions.

For example, there is a reaction involving the minerals garnet, **cordierite**, quartz and any one of the three aluminosilicates (Al_2SiO_5):

$$\text{garnet(s)} + \text{quartz(s)} + Al_2SiO_5\text{(s)} \longrightarrow \text{cordierite(s)} \qquad (8.4)$$

Cordierite is a silicate mineral found in some aluminous metamorphic rocks. Its composition varies from $Mg_2Al_4Si_5O_{18}$ to $Fe_2Al_4Si_5O_{18}$. Similarly, garnet in such rocks has a composition that varies from $Mg_3Al_2(SiO_4)_3$ to $Fe_3Al_2(SiO_4)_3$. The position of the phase boundary of reaction 8.4 depends on the precise Fe/Mg ratio of the rock. In Figure 8.5, this phase boundary is plotted for both the iron-rich compositions and the magnesium-rich compositions determined from experiments containing only Fe and only Mg compositions respectively. For real rocks containing both Fe and Mg, the reaction will lie somewhere between these boundaries, depending on the Fe/Mg ratio of the rock. For example, Figure 8.5 also shows the phase boundary for a natural cordierite with Fe/Mg = 1.5. For reactions involving such minerals, the geologist must not only identify which minerals are present but also analyse the precise composition of the cordierite

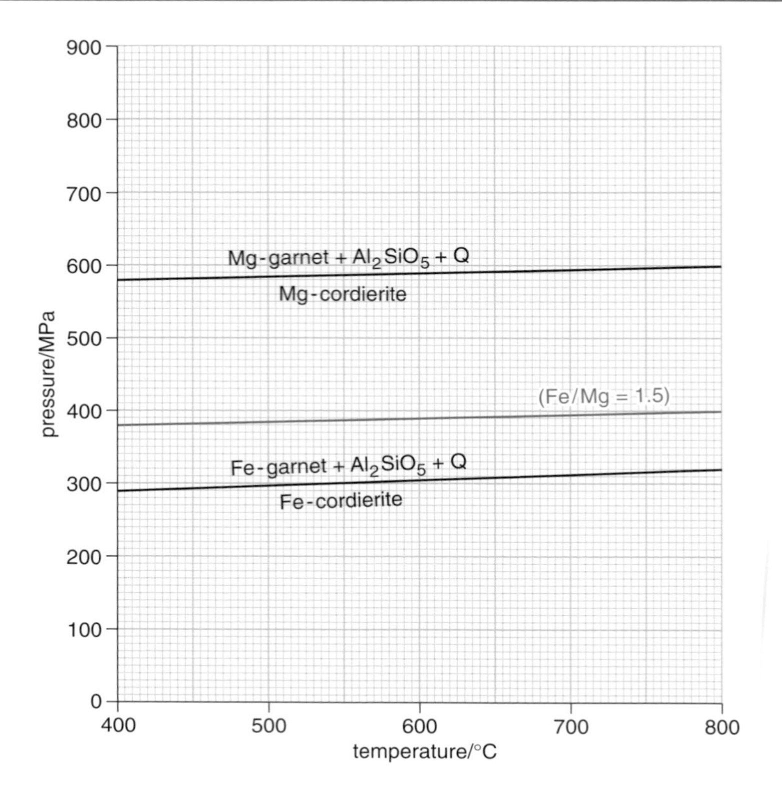

Figure 8.5 Phase boundaries for reaction 8.4 using Fe and Mg end-members. Also shown (red line) is the phase boundary for a typical cordierite with (Fe/Mg) = 1.5.

and garnet to find the appropriate position of the phase boundary at which the two minerals can coexist with an aluminosilicate and quartz.

❑ What is the main difference between the phase boundaries shown on Figure 8.5 and that on Figure 8.3?

■ The boundary on Figure 8.3 is much steeper.

❑ Which reaction gives a better idea of the pressure at which the phases on either side of the boundary can coexist?

■ The one involving cordierite, garnet, sillimanite and quartz gives a narrow range of pressures for a given Fe/Mg ratio.

❑ If a rock contains negligible Mg, what is the pressure range over which the mineral cordierite, garnet, sillimanite and quartz can coexist?

■ From the lower boundary in Figure 8.5, a range of 290–320 MPa is indicated.

The reason that this reaction is nearly parallel to the temperature axis is that it involves a large change in molar volume (ΔV) but only a small change in entropy. The change in volume is because garnet is a dense phase (density = $3.58 \times 10^3 - 4.32 \times 10^3$ kg m^{-3}) compared to cordierite ($2.53 \times 10^3 - 2.78 \times 10^3$ kg m^{-3}). Reactions with a gentle slope as described by Equation 8.4 make good natural barometers and those with a steep slope, like Equation 8.1, make good natural thermometers of metamorphic conditions. Good natural barometers are provided by reactions with a large change in molar volume (ΔV) (but should not also involve a large change in entropy), and good natural thermometers are provided by reactions with a large change in entropy (ΔS).

8.3 TYPES OF METAMORPHISM

At the beginning of this Section, we described metamorphic processes as those that are associated with mineralogical and structural changes in rocks. We have seen a few of the sorts of reactions that take place, the factors that control them, and the methods used to estimate the conditions of temperature and pressure. We shall now look a little more closely at the kinds of rocks that metamorphism produces, and the three main types of metamorphism.

8.3.1 DYNAMIC METAMORPHISM

Dynamic metamorphism takes place in areas of intense *local* deformation such as in fault zones or shear zones. Temperatures and pressures may be low if dynamic metamorphism occurs at high levels in the Earth's crust where rocks are brittle, but at deeper levels, higher temperatures result in more 'ductile' structures in which flow predominates over fracture. The differences between brittle and ductile deformation will be examined more fully in Section 9.

In a major fault zone, there is mechanical movement of one rock over another so that any rocks within the fault zone may be ground down in a process known as **cataclasis** (from the Greek meaning 'break down'). The resulting rocks are called cataclastic rocks. Almost all brittle faults contain a zone in which there are cataclastically broken or crushed rocks known as a **fault breccia** (Plate 8.2).

More ductile deformation causes a slippage between layers within the rocks, and between planes of atoms within minerals. Recrystallization under these conditions results in a fine-grained foliated rock with a streaked-out texture indicating the direction of shearing, which is known as a **mylonite** (Plate 8.3a). During the formation of mylonites, quartz is broken down into fine-grained aggregates that form elongated ribbons (Plate 8.3b). Note that the observation made from Activity 8.1 that grain size increases with metamorphic grade is not the case during mylonite formation or indeed any kind of dynamic metamorphism because dynamic metamorphism tends to reduce grain size.

Movements along fault planes expend considerable amounts of energy to overcome friction and most of this energy is released as *heat*. Although the size of the resultant increase in temperature depends on (i) frictional heating along the fault plane and (ii) the thermal conductivities of rocks on either side, it appears most sensitive to (iii) the rate of movement of the fault. Earthquakes are produced by rapid movements on faults and it can be shown that movements of this kind, although of very short duration, can produce a temperature rise sufficient to cause melting in the deformed rock – 5 cm per second has been suggested as the sort of speed at which rock melting might be expected.

Rocks melted by frictional heat are rare, but they do turn up in small quantities all over the world. Such melts form black, fine-grained rocks that look like basaltic glass and because they intrude the country rocks adjacent to the fault zone in small irregular dykes and veins (Plate 8.4a), they have often been identified wrongly as igneous rocks of more conventional origin, hence their name **pseudotachylites** (tachylite being an old term for basaltic glass). Many of the most spectacular pseudotachylites are now known to result from meteorite impact (Plate 8.4b). Following the impact, rapidly formed extensional faults allow rock masses to collapse towards the crater; the pseudotachylite is formed along these faults. This is a rare but impressive example of dynamic metamorphism.

8.3.2 CONTACT METAMORPHISM

Contact metamorphism (also known as thermal metamorphism) refers simply to the changes taking place in response to heat associated with igneous bodies (Block 2 Section 8.1). It is most obvious around intrusive rocks, because, unlike their extrusive equivalents, most of their heat is not lost to the atmosphere but is dissipated into the surrounding country rocks. Naturally, the temperatures are highest close to the igneous body itself, and there is therefore a very marked increase in metamorphic grade near the contact. The zone of metamorphic rocks around the intrusion is termed a metamorphic aureole (Block 2 Section 8.1).

Since the emplacement of magma rarely involves significant deformation (at least at shallow crustal levels), the main characteristic of rocks that have undergone contact metamorphism is the formation of metamorphic minerals in the absence of a tectonic fabric (cleavage, schistosity, or lineation) forming a hard, compact and splintery rock called a hornfels. Frequently, the first sign of contact metamorphism in the metamorphic aureole is not distinct new individual minerals, but a kind of 'spotting' caused by the growth of clusters of new metamorphic minerals. Such rocks are called spotted hornfels.

In the previous Section on metamorphic reactions, we stressed the importance of pressure and temperature in determining which minerals are likely to be stable (Figures 8.2–8.5).

❑ What is the third, and arguably most important, factor that ultimately governs whether a particular metamorphic mineral will be stable?

■ It is the bulk *composition* of the rock being metamorphosed: rocks of different chemical composition tend to contain different minerals, even when they have been at equilibrium under similar conditions of temperature and pressure (Block 2 Section 8.3).

The reason for the importance of chemical composition is simple. If a rock does not contain the appropriate elements for a particular mineral, then that mineral cannot be formed. Make sure that you recall the gist of Block 2 Section 8.4 by attempting Question 8.3 (you may wish to refer to Block 2 Table 7.2 and Figure 8.3).

Question 8.3 Which of the minerals garnet, calcite, muscovite and feldspar are likely to be present in the metamorphosed equivalents of the following rocks? Explain your answer in each case.

(a) Basalts.

(b) Limestones.

(c) Mudstones.

In a metamorphic aureole, high temperature minerals will grow nearest the contact and low temperature minerals are found furthest away. We can plot on a map where these metamorphic minerals occur in just the same way as we can map any other features, and in so doing would map **metamorphic zones**. Each zone is known by the name of a particular characteristic mineral present in that zone known as an **index mineral**. For example, we may speak of the biotite zone or the garnet zone. Provided we are looking at rocks of similar composition, the index mineral is a qualitative indication of the metamorphic grade. In practice, this technique is extremely useful where rocks of similar compositions exist over a large enough area, since the metamorphic zones then provide an indication of the relative changes in metamorphic grade, even when precise estimates of pressure and temperature are not available.

For example, the aluminosilicate polymorph that occurs very close to the contact of the granites of south-west England is andalusite. The distribution of andalusite around a granite defines the boundaries of the *andalusite zone*. Further away from the granite, where the temperature of the rocks reached during intrusion was lower, andalusite is not found, but other minerals will be present that grew at lower temperatures.

In this way, a series of metamorphic zones can be built up around an intrusion, each zone representing successively lower temperatures away from the body as, for example, has been done around the Ardara granodiorite in Donegal (Figure 8.6). In this example, the two aluminosilicates andalusite and sillimanite are particularly useful.

❑ By reference to Figure 8.2, can you determine whether, for any particular pressure, sillimanite is stable at higher or lower temperatures than andalusite? Is that consistent with their relative positions in the contact aureole in Figure 8.6?

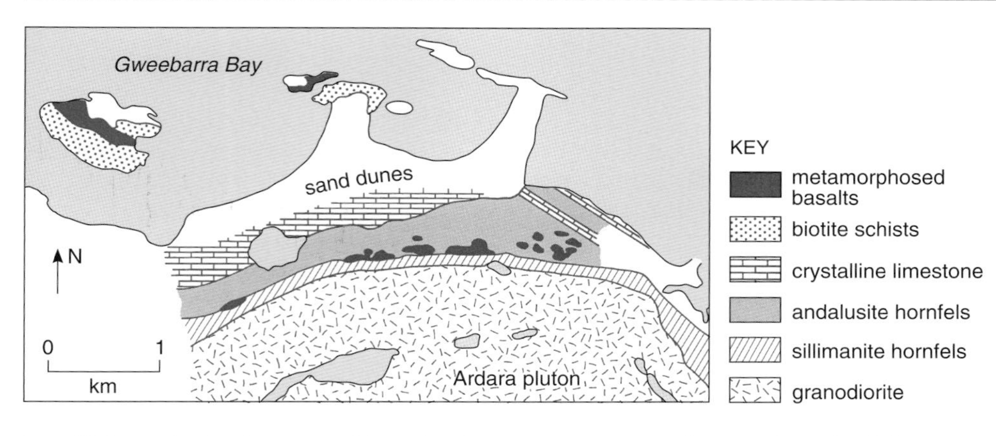

Figure 8.6 Metamorphic aureole around the Ardara pluton, Donegal.

- Sillimanite is always stable at higher temperatures than andalusite (Figure 8.2), and so occurs nearer to the granite contact than andalusite in Figure 8.6.

Question 8.4 (a) Since andalusite rather than kyanite reacts to form sillimanite in the contact aureole of the Ardara pluton, what is the maximum pressure at which metamorphism could have taken place?

(b) Assuming the pressure at which the granodiorite was emplaced to form the metamorphic aureole in Figure 8.6 was 200 MPa, use the phase diagram (Figure 8.2) to estimate the temperature at the outer margin of the sillimanite zone.

In general, large-scale changes in the bulk chemical composition of original rocks do not occur during metamorphism. This applies to contact metamorphism, but with some important exceptions. Right up against the igneous contact, carbonate-rich rocks are sometimes replaced by calcium and magnesium silicates, iron oxides and even sulphides. In this case, these elements have been introduced from the crystallizing magma.

- ❑ What do we call the changes of bulk composition that can occur during heating of rocks?
- Metasomatism. This was described in Section 5.1 associated with hydrothermal fluids.

In some metamorphic aureoles, metasomatism has been of direct practical use since it has resulted in the concentration of economic amounts of ores of elements such as lead and zinc.

Activity 8.2

We would like you to go on a virtual field trip to a metamorphic aureole in Cumbria. This is set up for you in Activity 8.2, which you should perform at the next convenient opportunity.

8.3.3 Regional metamorphism

Regional metamorphism is a very much larger-scale phenomenon than contact metamorphism (Block 2 Section 8.2). Regional metamorphism is usually associated with continental collision or subduction, and therefore with deformation. At the highest grades, partial melting of rocks occurs and ultimately granites are formed.

The patterns of changing metamorphic grade differ between contact and regional metamorphism. In contact metamorphism, the zones tend to be concentric around a particular intrusion, whereas in areas of regional metamorphism, the zones are often linear – as you might expect if they represent deeply eroded mountain belts, perhaps formed along some ancient destructive plate boundary or collision zone.

The third and perhaps most important distinction between contact and regional metamorphism is that the sources of heat are different, which has important implications for the rate at which metamorphism occurs. During contact metamorphism, the heat source is a cooling igneous body. This is an effective heat source only for as long as the magma is crystallizing. Even for a large granite body, this will be less than 100 000 years. For a small dyke, heat may be exhausted after 100 years or less. During regional metamorphism, the heat source comes largely from radiogenic heat from the decay of heat-producing elements. This is a much slower process. For example, garnets from metamorphosed sedimentary rock in the Himalayas have grown during regional metamorphism following continental collision. Plate 8.1 shows one such garnet. Highly precise dating of similar garnets has shown that although the core of a crystal formed 30 Ma ago, the outer parts of the garnet are only 25 Ma old. This result indicates that during regional metamorphism rocks can be heated for five million years at least.

Our present understanding of the metamorphism at continental collision zones is possible only because of the foundations laid by a handful of 19th century geologists. In 1893, George Barrow made a classic study of the regionally metamorphosed rocks of the Scottish Highlands where a sequence of sedimentary and volcanic rocks, some 10–20 km thick, was highly deformed and metamorphosed during what we now know to have been a period of subduction and collision about 500 Ma ago. Barrow mapped these rocks, and observed that provided he considered only the metamorphosed mudstones and shales, often called **pelitic rocks** (or pelites), there was a regular distribution of metamorphic minerals. Moreover, he was able to establish a series of zones, each characterized by a distinctive mineral (Figure 8.7).

Barrow was the first person to recognize metamorphic zones and to use index minerals to identify successive grades of metamorphism from the appearance of distinctive new minerals in a traverse across rocks of successively higher grade. Each new mineral represents a further step towards higher temperatures

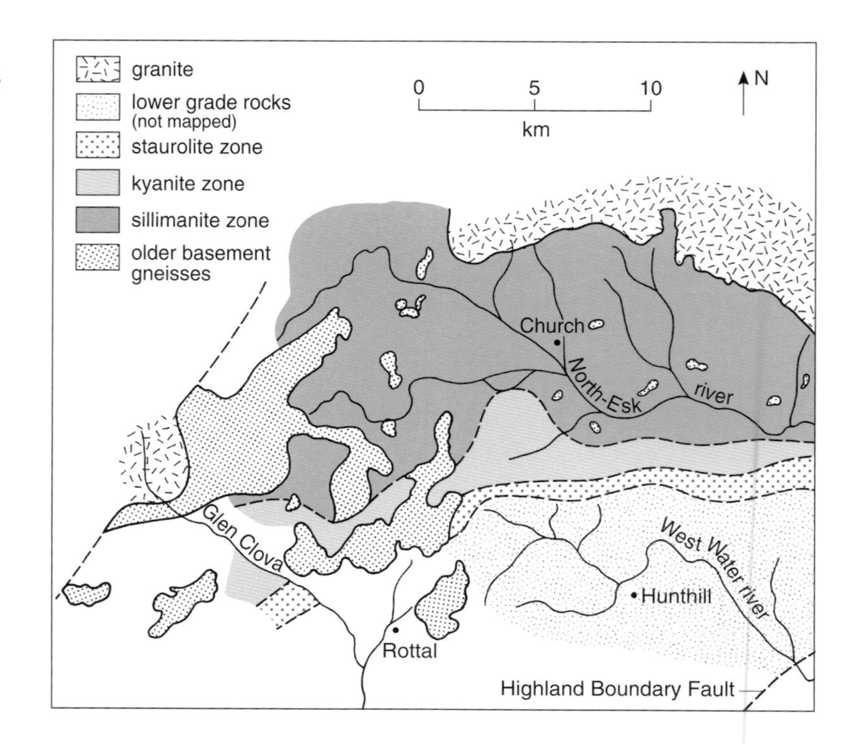

Figure 8.7 Simplified map of Barrow's original metamorphic zones in the pelitic rocks of the south-east Highlands of Scotland.

and/or pressures, and so becomes an index indicative of metamorphic grade. He recognized in the metamorphic rocks of the south-eastern Highlands six metamorphic zones and the index minerals he based them on were (in order of increasing metamorphic grade) chlorite, biotite, garnet, staurolite (a brown iron–aluminium silicate), kyanite and sillimanite.

What can we say about the conditions of pressure and temperature reflected in the appearance of these index minerals? Once again, the approach is very simple in principle – we do experiments at known pressures and temperatures and observe the reactions that take place. For example, we can interpret the change in P–T conditions between the kyanite and sillimanite zones by considering the experimentally determined phase diagram in Figure 8.2.

❑ How do the aluminosilicates recognized by Barrow compare with these present in the metamorphic aureole in Figure 8.6?

■ In the metamorphic aureole, there was andalusite and sillimanite, whereas in these Scottish rocks we find kyanite and sillimanite.

❑ Do the minerals in the Scottish examples reflect higher or lower pressures than those in the metamorphic aureole?

■ In Figure 8.2, the phase boundary between kyanite and sillimanite exists only at pressures greater than 370 MPa, whereas that between andalusite and sillimanite is at less than 370 MPa. Thus, the transition from kyanite to sillimanite in the Scottish rocks indicates that they were formed at higher pressures (>370 MPa).

Question 8.5 Figure 8.8 is the phase diagram for a reaction in which garnet is formed. Having argued that the pressure in this area was greater than 370 MPa, use Figure 8.8 to infer the approximate temperature at which garnet first appears, that is, at the beginning of the garnet zone.

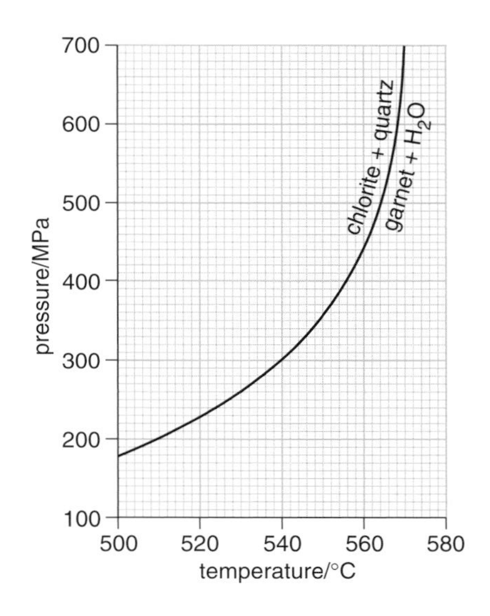

Figure 8.8 Phase diagram for the reaction chlorite(s) + quartz(s) $\rightleftharpoons$ garnet(s) + H_2O(g).

In the case of contact metamorphism, the near-concentric mineral zones around an igneous body essentially represent zones of successively higher temperatures. Although the same principles apply, regional metamorphism is more complicated since both pressure and temperature vary. In both types of metamorphism however, the points of first appearance of an index mineral can be mapped and linked by a line. Provided such minerals are *in rocks of similar composition*, this line marks the position of rocks with 'equal metamorphic grade' and so is known as an **isograd**. Conventionally, an isograd is named after the index mineral on the side of *higher* metamorphic grade. Thus, the border between the kyanite and sillimanite zones in Figure 8.7 is marked by the appearance of sillimanite as an index mineral and is called the sillimanite isograd.

It could be argued from Figure 8.2 that the transition from kyanite to sillimanite does not *necessarily* involve an increase in metamorphic grade, but could reflect a decrease in pressure. This possibility can be excluded by a more detailed examination of the minerals on either side of the isograd, and in general metamorphic zones are determined by changes of temperature rather than by changes of pressure.

Question 8.6 Figure 8.9 is a geological sketch map illustrating some folded metamorphosed sediments (metasediments) and a gabbro intrusion.

(a) Sketch in and label the isograds between the different mineral zones.

(b) Is the metamorphism older or young than the folding?

(c) Does the metamorphism appear to be related to the gabbro intrusion? That is, is it an example of contact metamorphism?

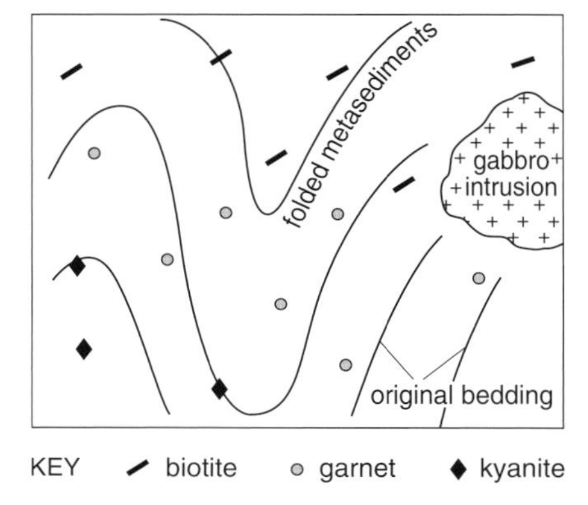

Figure 8.9 Geological sketch map, for use with Question 8.6.

The isograds you drew on Figure 8.9 may look convincing, but do not assume that they can be drawn as precisely as, say, a topographic contour line. Rocks have complex chemical compositions involving many elements and thus represent messy chemical systems; however, several experienced metamorphic geologists working in the same area would usually put an isograd at roughly the same place on the map. The significance of isograds for the rate of temperature increase with depth, or **geothermal gradient**, during metamorphism may not be clear cut. Unfortunately, in going from south to north across Barrow's zones, you are going from an area of low-grade metamorphism to one where both the temperatures and pressures were higher. Whether such variations represent a true geothermal gradient, in the sense that the minerals all crystallized simultaneously and so reflect variations in pressure and temperature *at one time*, or whether they crystallized at different times, is a subject we shall return to later.

Finally in this Section, we must reiterate the importance of rock composition in determining which metamorphic minerals will crystallize. If we have a sequence of pelites, carbonate-rich, and mafic rocks, different minerals will grow in each rock type and so there will be different *index* minerals in each rock type. We have seen that garnet, staurolite, kyanite and sillimanite can be used as important index minerals to establish metamorphic zones; they are all rich in aluminium, and they occur in Al-rich metamorphosed sediments (pelites). There are fewer mineralogical changes that can occur in carbonates and mafic igneous rocks (see Table 8.1 and also Block 2 Figure 8.3) and so pelites are generally more useful in establishing metamorphic zones. However, it is important to realize that if, as is common, metamorphic zones are based on index minerals found in pelites, the zones may well *include* rocks of other compositions. So there is nothing wrong with talking about mafic igneous rocks being in the staurolite zone, even though they contain no staurolite.

Table 8.1 Metamorphic assemblages produced in pelitic, carbonate-rich, and mafic igneous rocks.

	Assemblage produced (index mineral in italics):		
Metamorphic zone (based on pelites)	**Pelites**	**Carbonate-rich rocks**	**Mafic igneous rocks**
garnet	*garnet*, mica, quartz, Na-rich plagioclase	garnet, epidote, amphibole	chlorite, albite, epidote
staurolite	*staurolite*, garnet, quartz, mica, Na-rich plagioclase	garnet, anorthite, amphibole	amphibole, plagioclase
kyanite	*kyanite*, garnet, quartz, mica, Na-rich plagioclase	garnet, anorthite, amphibole	amphibole, plagioclase
sillimanite	*sillimanite*, garnet, quartz, mica, Na-rich plagioclase	garnet, pyroxene, anorthite	amphibole, plagioclase, garnet

Question 8.7 Figure 8.10 illustrates a series of isograds based on minerals in pelitic rocks. The isograds cut across a series of carbonate-rich, pelitic and mafic igneous rocks. Use Table 8.1 to indicate what mineral assemblages you would expect to find at points A, B, C and D.

8.4 METAMORPHIC FACIES

The mapping of metamorphic zones using index minerals is very useful for describing metamorphism from a particular area. However, it is difficult to compare the metamorphic grades of different areas on the basis of index

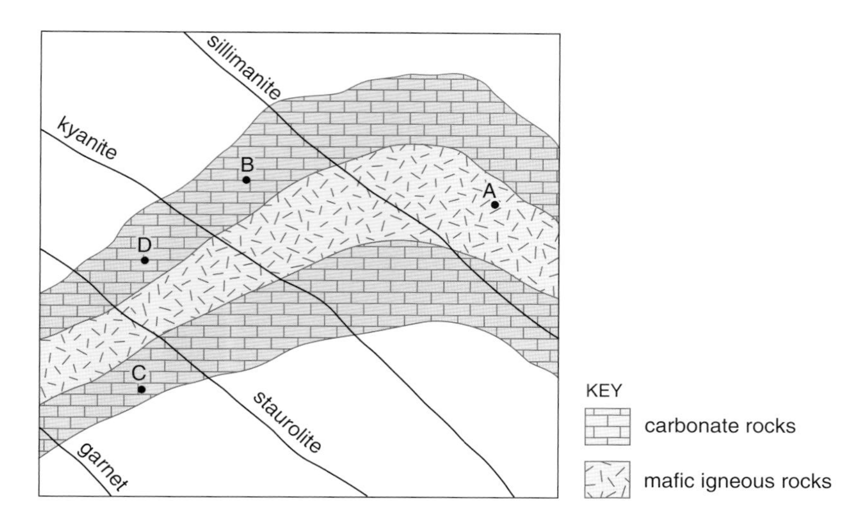

Figure 8.10 Isograds (based on pelitic rock minerals) cross-cutting a sequence of mafic igneous and carbonate-rich rock types. (For use with Question 8.7.)

minerals because of the possibility of different bulk compositions in the two areas, which will result in different metamorphic index minerals even under similar conditions of pressure and temperature. **Metamorphic facies** is a term used to embrace *all* the possible metamorphic mineral assemblages produced in rocks of different composition at similar temperatures and pressures. The assignment of a rock to a metamorphic facies is based on the observed mineral assemblage, which corresponds to a particular P–T range of metamorphism.

This definition requires amplification:

1 The composition of a metamorphic rock determines its mineral assemblage under any particular conditions of temperature, pressure and fluid composition. Thus, given a chemical analysis, it should be possible to predict the mineral assemblage that is likely to be stable under a given set of physical conditions (see also Table 8.1 and the discussion in Block 2, Section 8.4).

2 A metamorphic facies is determined from the *observed mineral assemblages* in associated rocks: it does *not* require knowledge of the precise conditions of temperature and pressure, which are estimated subsequently from the petrogenetic grid.

The first point we have met before, but the second is important because it often causes confusion. We have seen something of how metamorphic conditions may be estimated from experimental results on particular metamorphic reactions. Moreover, when data on a number of reactions that occur under different conditions are plotted together on a P–T diagram, it forms what we have called a petrogenetic grid – a criss-cross of different phase boundaries that allows us to infer likely metamorphic conditions from particular mineral assemblages. But such a petrogenetic grid is obviously experimentally determined, and thus, although its general form is unlikely to change significantly, it will be continually modified as experimental techniques develop and more experiments are carried out, particularly in the presence of fluid phases of different compositions. A metamorphic facies, by contrast, is assigned to the rock from the observed mineral assemblages. Provisional limits of temperature and pressure may then be assigned to the particular metamorphic facies from the petrogenetic grid as currently understood.

Let us take an example that you have already met. In Barrow's area of the Scottish Highlands, pelitic rocks in the *higher grades of metamorphism* contain kyanite and sillimanite, whereas the carbonate rocks contain garnet and pyroxene. Although the minerals found in the two types of rock are different because the rocks have different chemical compositions, both assemblages belong to the same metamorphic facies. In areas of *lower grade* metamorphism, pelites contain chlorite and biotite, and the carbonate rocks contain calcite, epidote and an amphibole, both rock types belonging to the same metamorphic facies.

❑ What metamorphic zone would the mineral garnet indicate if found in a carbonate rock?

■ From Table 8.1, garnet is present in all metamorphic zones in carbonate rocks. It does not therefore help to define the particular zone.

❑ Will a garnet-bearing rock of pelitic composition always be from the same metamorphic facies as a garnet-bearing rock of a carbonate composition?

■ No. Because they can be from different metamorphic zones, they can also be formed at different pressures and temperatures. If so, the metamorphic facies will be different.

Today, we recognize eight principal metamorphic facies that cover temperatures from 100 °C to over 800 °C, and pressures from atmospheric pressure to over 1400 MPa (Table 8.2). One of these (eclogite) you have met before in Section 7.4. There it was defined as a particular rock type; a metamorphic rock of mafic composition with garnet and pyroxene present. Here we are using the term **eclogite facies** to indicate the *P–T* field over which such an assemblage is present (i.e. a metamorphic facies). Although it is not necessary to define each facies here (definitions are given in the *Glossary*), the facies names generally result from the stable assemblage found in rocks of mafic composition. For example, **blueschist facies** rocks contain blue minerals only if the bulk composition is mafic. A pelite, when metamorphosed under blueschist-facies conditions, will be made up largely of white mica and quartz, but neither of these are blue! Nonetheless, it would be correct to say it is a blueschist-facies rock.

Table 8.2 Summary of metamorphic facies and their likely environments of formation.

Facies	Environment
zeolite	deeply buried sediments
prehnite–pumpellyite	more deeply buried sediments
blueschist	relatively high pressure but very low temperature
greenschist	extremely widespread facies, low-grade regional metamorphism in orogenic areas
amphibolite	also extremely abundant: high-grade regional metamorphism in orogenic areas
granulite	highest grade of regional metamorphism, occurs in the lower crust and within some Precambrian cratons
hornfels	metamorphic aureoles: moderate to high temperatures but at low pressures
eclogite	very high pressures and moderate to high temperatures

Our present estimates of the *P–T* fields that characterize the eight metamorphic facies are reproduced in Figure 8.11. Broadly, we may link metamorphic facies to estimates of pressure and temperature – and so to the change in temperature with depth known as the geothermal gradient. For example, **hornfels** results from contact metamorphism which is characterized by high but localized geothermal gradients due to the associated magmatic activity. Very low geothermal gradients occur when cold material from the Earth's surface is taken down to lower crust or upper mantle depths. In between, there are more 'normal' geothermal gradients which characterize much of both the Earth's continental crust and its metamorphic rocks.

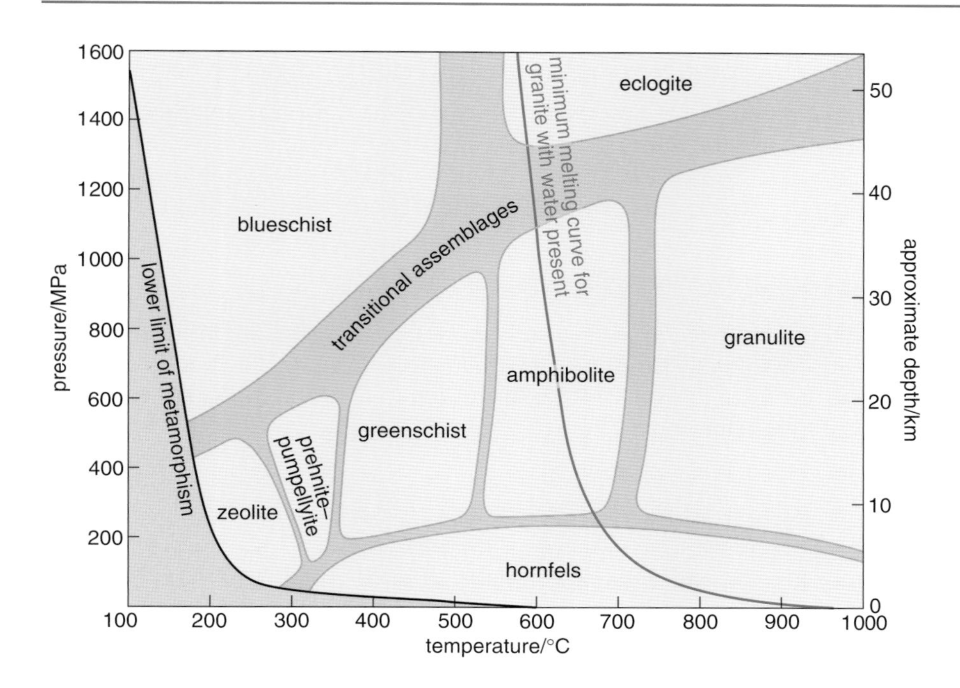

Figure 8.11 The *P–T* fields of the major metamorphic facies. (*Note*: In subducting oceanic crust, the transition of basalt to eclogite occurs, as stated on p. 59, at about twice the depth of the lowest-pressure part of the eclogite field on this diagram. The reason for this is that subducting oceanic crust is relatively cold (as implied by the isotherms on Figure 7.5 and overleaf on Figure 8.13). The geothermal gradient through subducting oceanic crust would therefore be a steep line that would cross from the blueschist field into the eclogite field some way above the top of the diagram.)

❑ If you plot a typical continental geothermal gradient of 30 °C/km on Figure 8.11, what sequence of metamorphic facies would you expect to find with increasing depth in the Earth's crust.

■ The geotherm starts near 0 °C at the surface and increases by 300 °C for every 10 km of depth. Near the surface, it passes through **zeolite** and **prehnite–pumpellyite facies**, then successively **greenschist** and **amphibolite facies**, reaching **granulite facies** at depth of about 25 km.

This observation raises some important questions. First, the occurrence of rocks indicative of metamorphic facies not in this sequence suggests that the Earth's geothermal gradient can be significantly disturbed. Secondly, how and why are rocks formed in higher pressure metamorphic facies subsequently brought back up to the surface? One explanation is that their exhumation results from isostasy in response to unusual thicknesses of continental crust. As an example, take the Barrow sillimanite isograd, which was formed at a pressure of 680 MPa, equivalent to a depth of ~25 km during metamorphism. At present, these rocks are exposed on top of a 35 km-thick crust. The simplest interpretation is that during the metamorphism that piece of crust was at least 25 + 35 km, or 60 km thick – compared to an average thickness for continental crust of 30–35 km. Since then, the thick crust has been uplifted, the top 25 km eroded and the sillimanite isograd has been exhumed and exposed at the surface.

❑ The metamorphic facies diagram covers a very wide range of temperatures and pressures. What other effects would you expect to see at very high temperatures?

■ Clearly, if you heat rocks to a high enough temperature, they are going to melt. On Figure 8.11, the red line that cuts across everything else represents the melting curve for rocks of granitic composition when water is present (it is the water-saturated solidus from Figure 4.3). At pressures greater than 400 MPa, melting takes place at about 600 °C, in the presence of H_2O.

Melts formed during regional metamorphism may rise to shallower levels and even cause their own metamorphic aureole. Thus, the effects of thermal and regional metamorphism may interact causing complex changes in metamorphic grade both horizontally and vertically.

8.5 Plate tectonics and metamorphism

The sequence of metamorphic facies recorded during deformation of the Earth's crust reflects the complex interaction between available heat sources and structural evolution. Some associations of metamorphic facies are indicative of particular tectonic settings and we shall examine two of these in a little detail.

8.5.1 Metamorphism and subduction

In general, the dominant heat source for regional metamorphism is supplied internally by decay of heat-producing elements (Section 2.4). The geothermal gradient (r) varies from one tectonic environment to another, but is related to the **heat flow** (q) by the equation:

$$q = Kr \tag{8.5}$$

where K is the **thermal conductivity** of the rocks. Heat flow is a measure of the amount of energy passing through unit surface area in unit time, usually expressed in milliwatts per square metre, mW m^{-2}.

Over much of the Earth's surface, the heat flow is surprisingly uniform at *c*. 60 mW m^{-2}. Striking exceptions occur near mid-ocean ridges and island arcs where the heat flow is high, and above oceanic trenches at destructive plate boundaries where it is low. This is illustrated in a profile of the heat flow variations observed across the western Pacific in Figure 8.12.

The low heat flow values on the flanks of the mid-ocean ridge in Figure 8.12 are due to cooling caused by the circulation of considerable volumes of seawater through the top few kilometres of the newly formed ocean crust.

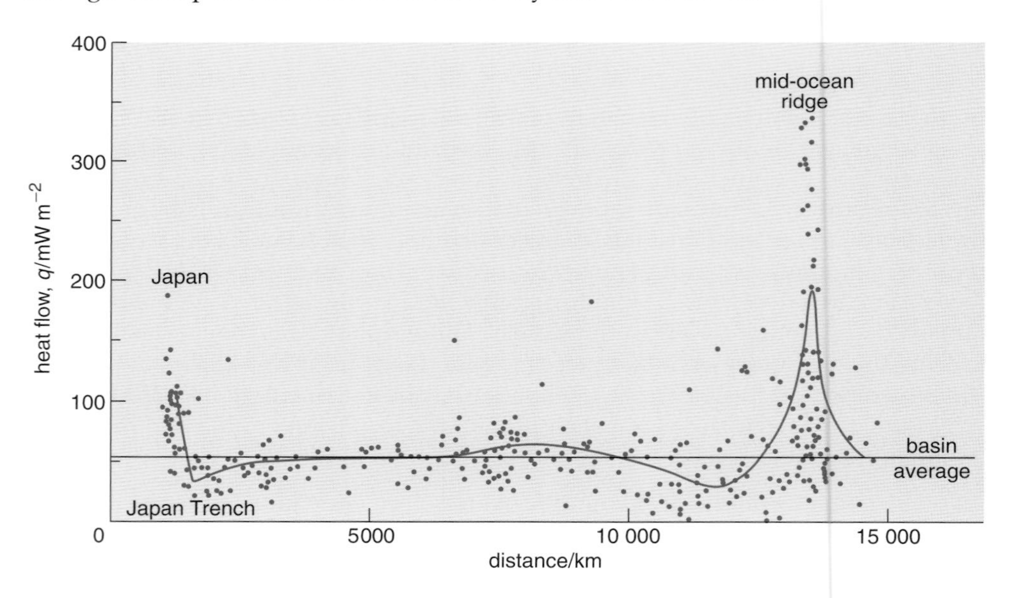

Figure 8.12 Variation in heat flow measurements across the western Pacific Ocean. The curved red line represents the average local heat flow, and the horizontal line illustrates the average heat flow from all ocean basins.

❑ Why are the heat flow values so much higher in Japan and on the mid-ocean ridge?

■ These are both areas where considerable volumes of magma are intruded. Magmas are hot, and as they move up towards the surface they bring heat with them, thereby raising the heat flow.

❑ Assuming that the conductivity does not change significantly, will the geothermal gradient be higher or lower in areas of higher heat flow?

■ Higher, since the geothermal gradient is proportional to the heat flow (Equation 8.5).

❑ What might be the cause of the low heat flow values in the Japan Trench?

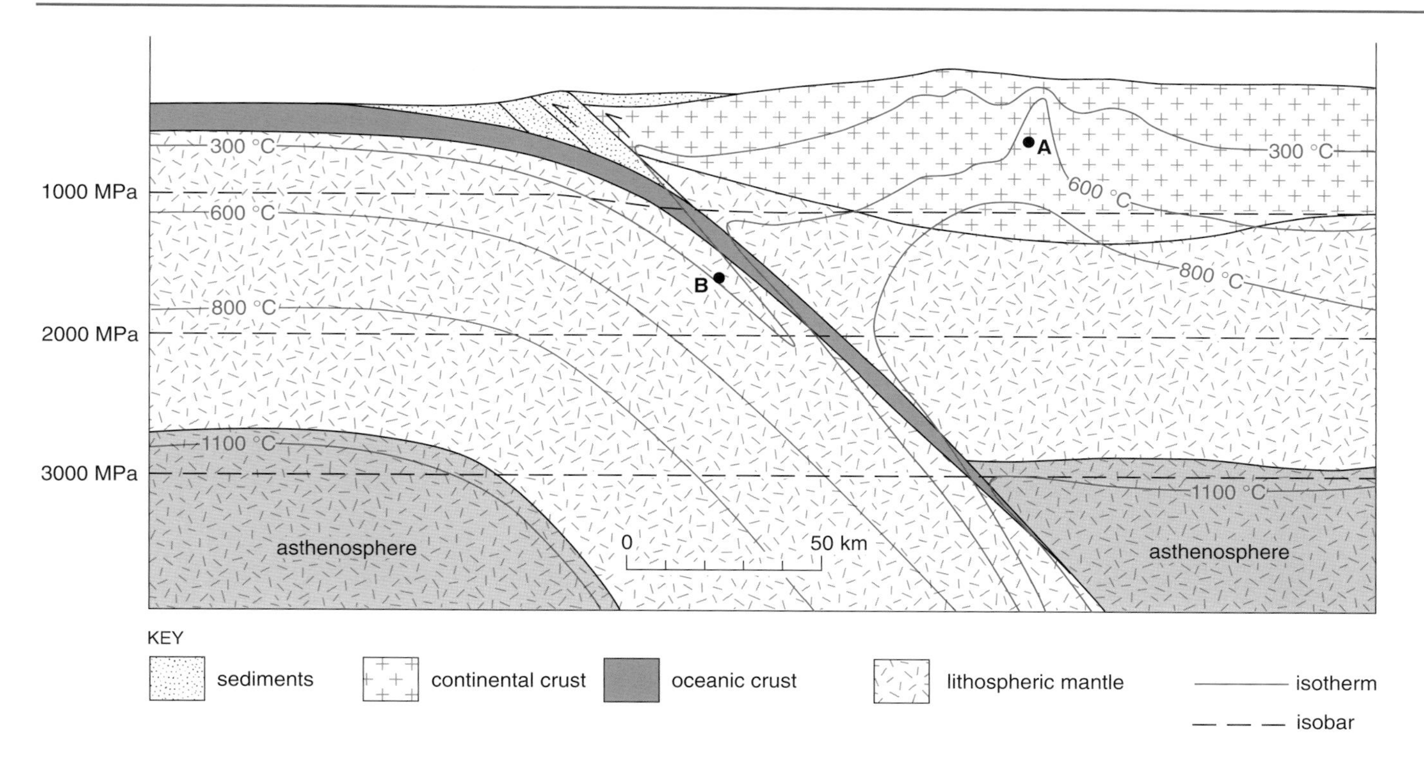

Figure 8.13 Cross-section through a destructive plate boundary illustrating high temperature–low pressure (A) and high pressure–low temperature (B) metamorphism. Together these make up a paired metamorphic belt, which runs parallel to the plate boundary.

- Such trenches mark the site of active subduction. The ocean crust has been at or near the surface. It is therefore relatively cold, and as oceanic lithosphere is subducted into the mantle the isotherms also are dragged down resulting in a low geothermal gradient and thus low heat flow to the surface (Figure 7.5, repeated in Figure 8.13).

In general, magmatic provinces are characterized by high heat flow and high geothermal gradients. Moreover, destructive plate boundaries would appear to be unique in that they contain a zone of high geothermal gradients (where the magmas are generated) adjacent to one where the temperatures have been reduced by subduction (Figure 8.13). The net result is that in addition to the distinctive igneous rocks (cf. Section 7.4) there should also be two roughly parallel suites of metamorphic rocks, one containing relatively high-temperature and low-pressure mineral assemblages and the other those typical of relatively high pressures and low temperatures. This is called a **paired metamorphic belt** and is characteristic of destructive plate boundaries.

Question 8.8 What sort of geothermal gradient and what metamorphic facies will be present at points A and B on Figure 8.13?

Paired metamorphic belts identified from around the Pacific were first explained simply in terms of the varying geothermal gradients that are associated with subduction zones (Figure 8.13). However, the presence of blueschist assemblages (B) requires not only that very high pressures occurred at low temperatures during active subduction, but also that the rocks were then uplifted rapidly to the surface before they had time to warm up. For this reason, blueschist-facies rocks are not always preserved in ancient subduction zones. In some examples from subduction zones, the initial low temperature/high pressure minerals warm and recrystallize as the rocks are brought to the surface resulting in the loss of blueschist-facies minerals and the preservation only of higher-grade metamorphic facies. Throughout the metamorphic history of any rock in a subduction (or collision) zone, there is a competition between the rate of heating imposed largely by radioactive decay of heat-producing elements and the rate of cooling imposed by the rock being brought towards the surface during exhumation. In the case of blueschists, their high P–low T assemblage is preserved only when the rate of exhumation (controlled largely by isostasy) exceeds the rate of heating (controlled by thermal conductivity of the rocks).

8.5.2 METAMORPHISM AND COLLISION

The destruction of an oceanic basin by complete subduction of the oceanic crust that formerly separated two continents is likely to be followed by continental collision and crustal thickening, as you learned from Section 7.5. Such momentous events lead to metamorphism on a regional scale.

- ❑ What is the heat source responsible for regional metamorphism during crustal thickening?
- ■ Most of the heat within the Earth is generated by the radioactive decay of U, Th and K, and these elements are most abundant in the rocks of the continental crust, particularly the uppermost crust. So a thicker crust means a steeper geothermal gradient.

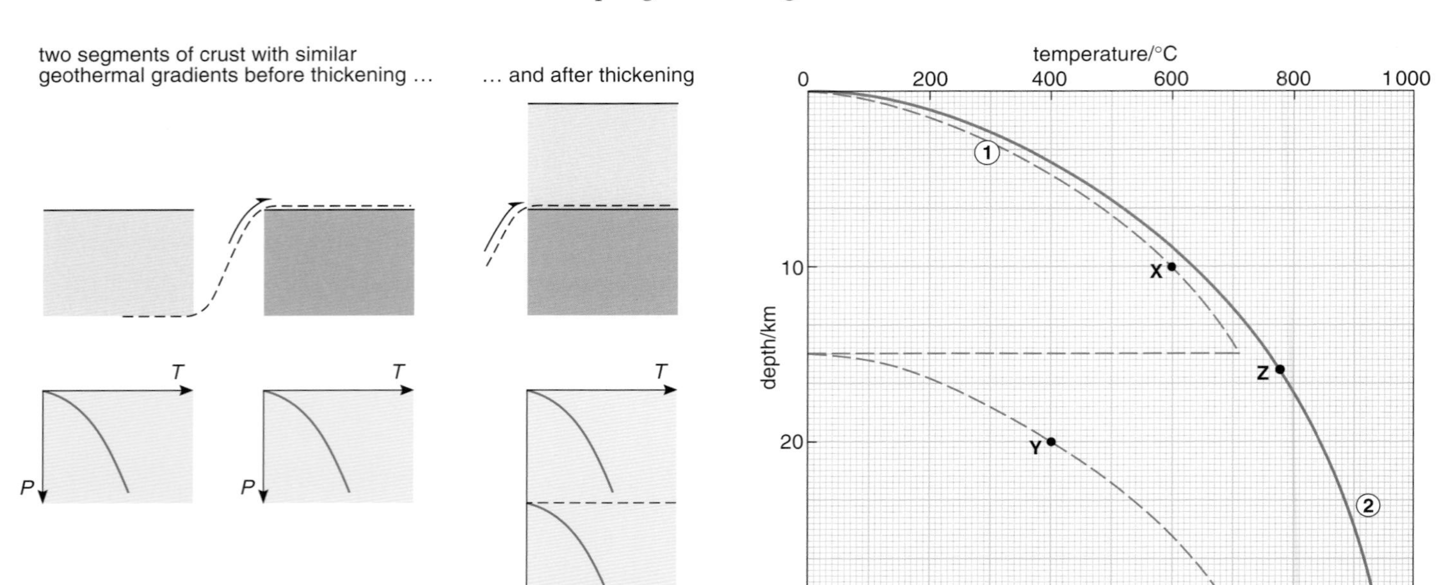

Figure 8.14 (a) Two segments of crust with similar geothermal gradients (shown as curved red lines) before and immediately after crustal thickening. (b) Temperature variation with depth: curve 1 (dashed), immediately after thrusting: curve 2, (solid) after 30 Ma. Points X, Y and Z are discussed in the text.

The changing geothermal gradient during continental collision can best be understood by considering the effects of thrusting one crustal slice over the other, hence thickening the crust. If we start with two segments of upper continental crust with the same geothermal gradient, the thrusting of one onto the other results in a distinctive 'saw-tooth' variation of temperature with depth (Figure 8.14a, and curve 1, Figure 8.14b). At this instant, the pressure in the lower segment has increased, but the temperatures remain as yet unchanged. This clearly cannot last since temperatures of near 0 °C cannot be maintained in what is now well within a thickened segment of continental crust. Instead, the continuing radiogenic decay of U, Th and K, which are now present in larger quantities since there is more upper crustal material, gradually increases the temperature, until at equilibrium, after about 30 Ma its variation with depth is illustrated by curve 2 of Figure 8.14b. Predictably enough, a rock at a depth of 17 km on Figure 8.14b (near the top of the underlying piece of crust) has had its temperature increased considerably, from 200 °C to 800 °C. What is perhaps more surprising is that even in the upper segment of crust the geothermal gradient has also been increased by crustal thickening.

An actual collision zone is naturally more complex than this model implies. In particular, because of isostasy the thickened crust will be susceptible to exhumation, and rocks must cool as they approach the surface. In some cases, this cooling effect is rapid enough to prevent the geothermal gradient from reaching the equilibrium temperature gradient indicated as curve 2 in Figure 8.14b. If the rocks are exhumed very rapidly after thrusting, then the metamorphic facies appropriate to the *P–T* conditions X and Y along curve 1 of Figure 8.14b will be preserved.

❏ From Figure 8.11, what are the metamorphic facies of rocks metamorphosed at X and Y?

■ X lies in the amphibolite facies (600 °C at 10 km) and Y in the greenschist facies (400 °C at 20 km) (Figure 8.11).

Both greenschist and amphibolite facies are represented in the Barrow zones, but what is unusual in Figure 8.14 is that the colder rocks (Y) occur at greater depths than the hotter rocks (X). Such a temperature inversion has been observed in metamorphic rocks of the Himalayas and its preservation has been interpreted as evidence of very rapid exhumation.

If, as is more usual, the geothermal gradient reaches equilibrium along curve 2 of Figure 8.14b, rocks deep in the thickened crust (Z) will reach temperatures of 800 °C or more, well within the granulite facies. In this case, the diagnostic feature of the collision zones is that the temperatures will increase more rapidly with depth than in crust of normal thickness. Many of the granulite-facies rocks exposed today result from heating of a crust thickened by collision and subsequent exhumation of deep crustal levels by isostasy.

8.6 SUMMARY OF SECTION 8

- Metamorphism is the mineralogical response of a rock to imposed conditions of temperature and pressure that differ markedly from those of its origin.
- Broadly, the study of metamorphic rocks comprises: (a) evaluation of the conditions of temperature and pressure to which the rocks were subjected during metamorphism, in the light of experimental data; (b) consideration of the thermal evolution of an area and its constraints on possible tectonic environments.
- Metamorphic grade is defined in terms of the entropy change (ΔS) of the particular metamorphic reactions: increasing metamorphic grade involves an increase in entropy of the metamorphic system. In general, higher-grade rocks reflect higher temperatures. *Prograde* reactions are those which take place with increasing metamorphic grade, and *retrograde* reactions are those which take place with decreasing metamorphic grade. Rates of reaction increase exponentially with temperature, and so equilibrium is more likely to be achieved at higher grades of metamorphism. The presence of a fluid phase will also greatly increase the rates of reaction and its absence will inhibit retrograde reactions.
- The *petrogenetic grid* is the pattern of experimentally determined phase boundaries plotted on a phase diagram of temperature against pressure. In principle, once the mineral assemblage in a metamorphic rock has been identified, the petrogenetic grid may be used to estimate the conditions of pressure and temperature under which it crystallized.
- *Dynamic* metamorphism occurs within localized areas of intense deformation, as along fault or shear zones. *Contact* metamorphism refers to changes in response to heat in the vicinity of igneous rocks. Typically it is not associated with deformation, and the metamorphic grade decreases systematically away from the contact with igneous rocks. *Regional* metamorphism tends to occur over large areas, and as it is usually associated with collision or subduction zones, the rocks will have been deformed.

- The particular minerals observed in a metamorphic rock reflect its *composition*, in addition to the conditions of pressure and temperature and the nature of the fluid phase. Thus *different* minerals occur in rocks of the same metamorphic grade, but different compositions, and the term *metamorphic facies* is used to embrace all possible mineral assemblages in rocks of different composition that are thought to have crystallized under similar *P–T* conditions. A metamorphic facies is assigned from the observed mineral assemblages, and not from the inferred conditions of pressure and temperature which are estimated from the petrogenetic grid.
- The mineral assemblages in metamorphic rocks from different areas can reflect different variations in temperature and pressure and these vary depending on the thermal and hence tectonic environment. Geothermal gradients are steep in areas of magmatic activity (e.g., island arcs and mid-ocean ridges), and low at subduction zones. Continental collision and crustal thickening also results in an increased gradient, due primarily to the radioactive decay of U, Th and K within the crust. However, the preserved mineral assemblages depend not only on the geothermal gradient during metamorphism but also on the rate at which the rocks are brought to the surface.

8.7 OBJECTIVES FOR SECTION 8

Now you have completed this Section, you should be able to:

8.1 Carry out observations on hand specimens of metamorphic rocks, identify metamorphic minerals and textures and interpret these in terms of the processes involved.

8.2 Interpret simple phase diagrams to infer the conditions of temperature and/or pressure at which a particular metamorphic mineral, or assemblage, was at equilibrium.

8.3 Map out isograds and use them to assess the relationship of metamorphism to both the intrusion of an igneous magma and a deformation event.

8.4 Evaluate estimates of metamorphic pressure and temperature from metamorphic assemblages to suggest possible tectonic environments in which the metamorphism could have taken place.

Now try the following questions to test your understanding of Section 8.

Question 8.9 A fairly common metamorphic reaction is that of

$$\text{staurolite(s)} + \text{quartz(s)} \longrightarrow \text{garnet(s)} + \text{sillimanite(s)} + H_2O(g)$$

Predict which assemblage has the lower entropy, which is stable at higher grades of metamorphism, and whether this reaction might take place at a staurolite or sillimanite isograd.

Question 8.10 A metamorphosed pelite contains the mineral staurolite in equilibrium with kyanite. Use the following information to estimate the approximate conditions of pressure and temperature during metamorphism.

(a) Staurolite does not form until the temperature exceeds 580 °C in pelitic rocks at pressures greater than 200 MPa.

(b) No granite rocks or migmatites were formed nearby, and yet pelite rocks melt at *c*. 650 °C under moderate pressures. You can assume H_2O was present during metamorphism.

(c) The phase relations in Figures 4.3 and 8.2.

Question 8.11 The inferred metamorphic conditions of temperature and pressure of two rocks from different areas are 200 °C at 800 MPa, and 550 °C at 200 MPa, respectively. For each rock, identify which metamorphic facies it is from, calculate the geothermal gradient (assuming that, for each 1 km in depth, pressure increases by 30 MPa) and suggest in what thermal or tectonic environment it might have been metamorphosed.

9 ROCK STRUCTURES

Discussion of continental collision and metamorphism prompts us to investigate how rock structures form. Structural geology is the study of **deformation**, the physical record left by processes that change the shape of rocks. In this respect, structural geology is like metamorphic geology; both branches aim to shed light on changes that affect rocks *after* they formed. If our aim is to understand the complete history of any rock, then it is vital to understand how structures form.

Deformation affects all types of rock – igneous, metamorphic and sedimentary – and can be observed on all scales. Structural changes are usually brought about by differential movement between large bodies of rock, commonly a combination of both horizontal and vertical movements. The features that provide evidence for such movement may be seen in exposures of rocks of all ages, all over the world. The five main structures we will consider are *joints*, *faults*, *folds*, *cleavage* and *schistosity*. To describe them, you will need to become familiar with some of the jargon of structural geology, but you have met many of these terms and concepts already in Blocks 1 and 2.

9.1 BRITTLE DEFORMATION AND JOINTS

One important way in which rocks can deform involves *fracture*. When a material fractures, it splits into discrete blocks which may or may not move relative to one another. This type of behaviour is called **brittle deformation**. A glass behaves in a brittle way when you drop it onto the floor. Individual pieces of glass (usually rather small) remain coherent and undamaged, but fractures separate the individual pieces which generally become displaced one from another across the floor. Brittle structures are common in rocks; the main brittle structures we see are *joints* and *faults*.

Joints are fractures in rocks which show *no appreciable lateral displacement* between the two sides of the fracture. They are probably the most common and often the most prominent structure visible in rock outcrops, irrespective of rock type (see Block 1 Sections 2.2–2.4). They usually take the form of more or less planar surfaces cutting right through the rock, although they can also be curved or irregularly shaped surfaces. Individual joints can frequently be traced for many metres. Joints commonly occur in groups of several parallel or almost-parallel fractures, known as **joint sets**. The spacing between adjacent joints in a set can vary from millimetres to several metres.

Rocks usually contain more than one joint set. In sedimentary rocks, joints can develop parallel to bedding, accentuating it, but you are more likely to notice joints oriented approximately at right angles to bedding (Figure 9.1, and Block 1 Figure 2.5). Commonly, joints intersect on the bedding surface in an X-like pattern, often called a **conjugate** pattern. In igneous rocks, joints occur both parallel to and at right angles to the surface of an intrusion; in metamorphic rocks joints often occur at right angles to a foliation.

❑ Look carefully at the joints in the sedimentary layers in Figure 9.1. How are the joint patterns in the prominent limestone beds different from those in the shale beds?

Figure 9.1 Well-bedded limestones and shales, with regularly spaced joints in the limestones approximately at right angles to bedding.

■ In the limestones, well-developed joints can be seen at right angles to bedding. These are absent from the shale beds, which appear to be unjointed.

Often where different rock types are interbedded, some lithologies are well-jointed, whilst the joints are less well developed in others. Joint development must therefore be dependent in some way on the physical properties of rocks within the succession.

Since joints are gaps and cracks in the rock, they form obvious fluid pathways. Where fluids flow through rocks, *veins* may be precipitated (Block 2 Sections 6.5 and 6.6). For this reason, joints seen at outcrop are often filled with thin seams of crystalline white mineral, which is often calcite or quartz.

9.2 FAULTS

Faults are discontinuities between rock masses which show a *measurable lateral displacement* between the two sides. Whilst joints may simply show movement apart between blocks of rock, faults are recognized by horizontal or vertical displacement between two blocks, or more typically a combination of both directions. The amount of displacement can vary from millimetres to many hundreds of kilometres. Faults with small displacements are much more common than those with large displacements.

To summarize what you should have gained from Activity 9.1, faults are usually described in terms of:

- the *dip* amount and the *strike* orientation (called the **trend**) of the *fault plane*, in degrees or compass directions;
- the *amount* of movement between the two sides of the fault (often simply the vertical component of movement, or *throw*), in metres or feet;
- the *sense* of this movement (combinations of the terms dip–slip or strike–slip, normal or reverse).

9.2.1 Hangingwalls and footwalls

The two displaced blocks of rocks separated by the fault plane are known as the **hangingwall** and the **footwall** (Figure 9.2b,c). These useful terms are old miners' names – as miners stood on an inclined lode working ore, they had their feet on the 'wall' underneath them whilst another block of rock 'hung' over their heads. The footwall is therefore the body of rock *underneath* the dipping fault plane, and the hangingwall is the body of rock *above* the fault plane. As long as the fault plane is not vertical, we can always recognize a hangingwall and a footwall. Either 'wall' may have been displaced during faulting; often both have. To remember which is which of these two important terms, think of someone standing on the fault plane – they have their feet on the footwall.

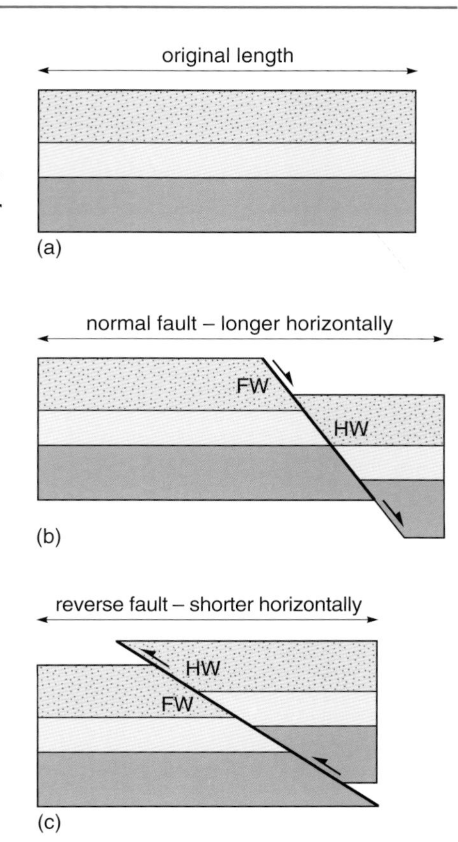

Figure 9.2 Cartoons that illustrate how normal faults allow rocks to lengthen horizontally, whilst reverse faults allow them to shorten horizontally. (a) Undeformed rock; (b) normal fault; (c) reverse fault. HW = hangingwall; FW = footwall. Half-arrows indicate the way the hangingwall has moved *relative to the footwall*.

These terms give us a better way of describing the movement sense of faults. In normal faults, the hangingwall has been displaced (relative to the footwall) *down* the dip of the fault plane, whereas in reverse faults the hangingwall has been displaced *up* the dip of the fault plane (Figure 9.2).

❑ Do normal faults allow a body of rock to lengthen or to shorten horizontally? How about reverse faults?

■ Since the hangingwall of a normal fault has been relatively displaced down the dipping fault plane, it must also have moved *outwards*. This means that the horizontal dimension must increase, so the body of rock must have lengthened horizontally (Figure 9.2b). This argument is reversed for reverse faults; upward displacement of the hangingwall must move it inwards, meaning that the body of rock has shortened horizontally (Figure 9.2c).

If a fault plane is exactly vertical, we cannot define a hangingwall or a footwall, so strictly we cannot classify a vertical fault as normal or reverse. In fact, most fault planes are actually inclined, so the problem seldom arises *as long as the fault plane can be seen*. However, you will come across situations where the fault cannot be seen directly, usually because surface debris hides the exact location of the fault. It can be difficult to be sure of the amount or sometimes even the direction of dip. In such cases, it is generally easiest to consider the fault to be vertical. By convention, we call vertical faults 'normal' unless there has been demonstrable strike–slip movement.

Strike–slip faults also have hangingwalls and footwalls (as long as the fault plane is not vertical) and the terms *dextral* and *sinistral* refer to the relative rightwards or leftwards displacement of *either* the hangingwall viewed from the footwall *or* the footwall viewed from the hangingwall. This movement sense is the same whichever way you look.

Activity 9.1

In Block 1, you learned a number of descriptive terms relating to faults which we will be using throughout the remainder of this Block. How many can you still remember? Try Activity 9.1 now to refresh your memory for fault terminology.

9.2.2 Slickensides

How can a geologist know which way the walls of a fault have actually moved? As you saw in Block 1, it is not generally possible to answer this question simply from a study of the rocks on either side of the fault. As long as we can distinguish between younger and older rocks, we can establish which wall has been downthown relative to the other. If in addition we know the dip of the fault, we can establish whether it is normal or reverse. However, none of this information enables us to establish the exact path one wall took relative to the other.

Figure 9.3 Aligned grooves (slickensides) parallel to the pencil on the surface of a small fault in the French Alps.

Clues to the movement direction lie along the fault plane itself. Many fault planes show a polished surface, called a **slickenside surface**, containing either aligned grooves or elongate crystalline fibres, commonly of quartz or calcite. These grooves or fibres are aligned in the movement direction of the fault (Figure 9.3). They developed when the fault was active; fibres form when the walls are held apart by mineral-rich fluid under pressure, grooves form where they are not. The grooves themselves are called **slickensides**, whilst the fibrous mineral growths are often called **slickenfibres**. Slickensides and slickenfibres give an accurate measure of the direction of slip of the fault. However, the movement history of the fault might be a protracted one, and generally speaking slickensides and slickenfibres record only the last (or one of the last) episodes of movement, so although they provide an accurate movement direction they rarely provide a *complete* record of movement. Grooves and fibres associated with real faults commonly show **oblique slip**, such as the ones in Figure 9.3. In such cases, relative displacement of the two walls is neither exactly up- or down-dip nor along strike, but somewhere in between.

9.2.3 Detachments

Faults usually cut across beds in sedimentary successions. However, a fault plane can lie parallel to beds, in its hangingwall or its footwall or both. Where this happens, the direction of downthrow relative to the dip of the fault cannot be worked out. Such flat-lying faults are neither normal nor reverse, and they are often called **detachment faults** (or simply *detachments*). Detachment surfaces are often only recognized as surfaces of displacement because they contain slickensides or slickenfibres.

9.2.4 Fault tip lines and fault arrays

Faults are three-dimensional structures, and fault planes extend laterally through rocks as well as penetrating upwards and downwards. They are never infinite, though. Faults show a maximum displacement at one point on the fault plane, and the amount of displacement decreases away from this point. At some distance, the displacement will die out altogether and so, therefore, does the fault. The locus of all the points on a fault plane where the **fault displacement** dies away to nothing is called the **fault tip line**. The point where the fault tip line intersects another surface, such as the ground surface or a displaced geological horizon, is called the **fault tip** (Figure 9.4a).

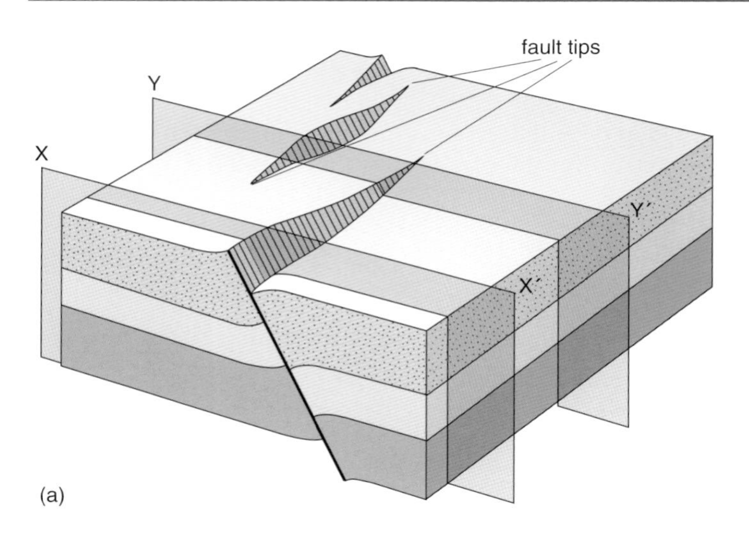

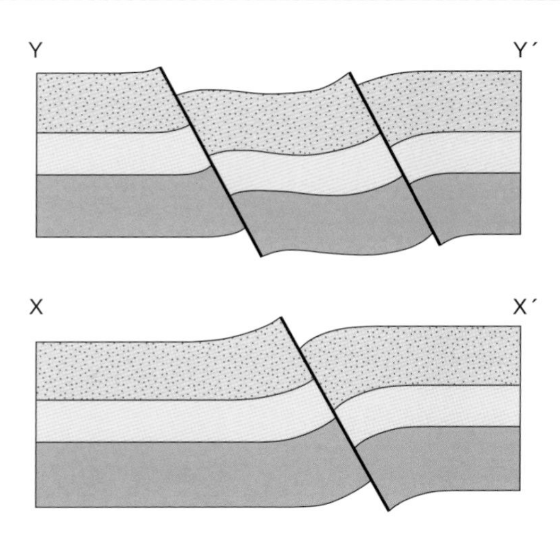

Figure 9.4 Fault tips and fault arrays. (a) A block diagram showing an array of normal faults, which together allow deformation of a large block of rock. Note how the individual faults overlap in plan view. The fault tip line intersects the displaced surface at fault tips. (b) Two cross-sections, X–X′ and Y–Y′, have been drawn through different parts of the block diagram. The number of faults in each cross-section is different, but the amount of horizontal displacement is the same (*Note:* The folds shown near to the faults are illustrative of those seen in the hangingwalls and footwalls of real major normal faults.)

Figure 9.4 shows schematically how faults of finite size develop together to allow displacement on a large scale. The displacement on each individual fault dies out, but displacement is picked up by a second fault which overlaps the first. In Figure 9.4a, there are three such overlapping normal faults. Even though each individual fault is only of finite length, the overall displacement is maintained across the body of rock, as can be seen from the uniform rectangular shape of the block. Multiple faults with parallel or sub-parallel strike directions are called **fault arrays**.

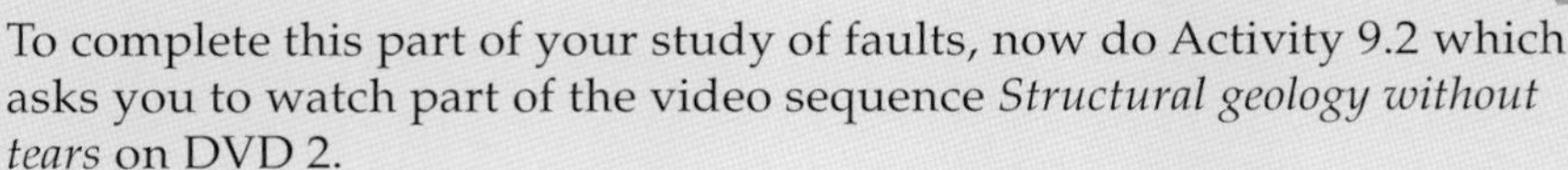

Activity 9.2

To complete this part of your study of faults, now do Activity 9.2 which asks you to watch part of the video sequence *Structural geology without tears* on DVD 2.

9.3 Shear zones and ductile deformation

Outcrops of metamorphic rock, in particular, sometimes show a structure known as a **shear zone** (Figure 9.5). Shear zones are fault-like structures in which there is evidence of displacement and deformation, but no brittle failure. In shear zones, displacement is distributed within the rock rather than taking place across a single plane.

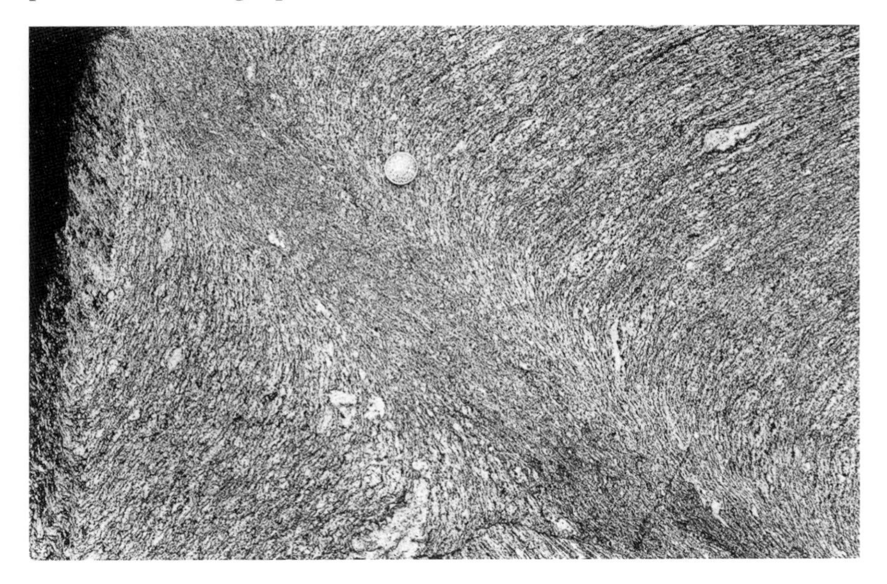

Figure 9.5 Photograph of a shear zone within deformed gneiss from Ticino, Switzerland. Note how the foliation in the rock sweeps anticlockwise into the zone, until it is almost parallel to the zone itself.

Deformation in which material is permanently deformed but does not fracture is called **ductile deformation**. Shear zones are therefore the ductile equivalents of brittle faults. Shear zones are much more common in metamorphic and plutonic igneous rocks which formed at depth than in sedimentary successions which formed at the surface, so in part they are can be considered as the deep-seated equivalent of faults.

9.4 Folds

Perhaps the most obvious ductile structures we see in rocks are *folded* rock layers.

Activity 9.3

In Block 1, you learned descriptive terms relating to folds which we need to use now. Try Activity 9.3 to remind yourself of these terms.

9.4.1 Describing folds

Folds are usually described in **fold pairs** (an adjacent anticline and a syncline taken together), and in terms of:

- The *dip* amount and the *strike* orientation (the *trend*) of the *fold axial surfaces*, in degrees and a compass direction.
- The amount and direction of *plunge* of the fold axes, also in degrees and a compass direction.
- If the fold pair is asymmetric, the *sense of asymmetry* of the fold, expressed as a direction of steepening or apparent overturning of the limbs of the fold. This is very often opposite to (i.e. 180° from) the direction of dip of the axial surface. For example, the asymmetric fold pair shown in Block 1 Figure 4.10 is asymmetric towards the south.
- The amount of the angle subtended internally between the two limbs, known as the **interlimb angle**, which may either be expressed in degrees or more usually by a simple descriptive term. Folds are described as being *gentle*, *open*, *close*, *tight* or **isoclinal**, if they have interlimb angles of 180°–120°, 120°–70°, 70°–30°, 30°–0°, and 0° respectively. The folds shown in Block 1 Figures 4.9 and 4.10 both have interlimb angles of 80° and would both be called *open* folds, whereas the anticline shown in Block 1 Figure 4.14 has an interlimb angle of 40° and is *close*.
- The general shape of the fold closure, either *rounded* like a 'U' or *angular* like a 'V'.

❑ How would you classify the folds in Plate 9.1, using some of the terms described above?

■ We cannot tell much about the exact dip and strike of the axial surfaces, or the exact direction and amount of plunge from this photograph. As far as we can tell, they appear to be symmetric. They have interlimb angles of between 60°–90°, so they must be open or close. They have straight limbs and relatively angular hinges, so they are angular rather than rounded. We might call them open, angular, symmetric folds. (Were we to visit this outcrop, we could measure dips and strikes and be more certain about the missing terminology.)

Deformed rocks show folds on all scales, so it is useful to be able to describe the dimensions of folds. The width of a fold pair is called the **fold wavelength**; the height of the fold is called the **fold amplitude**. These terms are analogous to other waveforms; the wavelength is the lateral distance between adjacent anticlinal (or synclinal) fold hinges and the amplitude is half the height between adjacent anticline and syncline hinges.

Question 9.1 Look at the folds in Plate 9.1 and roughly estimate (a) the wavelength and (b) the amplitude of these folds. (The fenceposts are about 1 m tall and the road is about 10 m wide.)

Folds are further classified using two terms, with reference to the dip of their axial planes (approximating the axial surface to a plane) and to their plunge (Figure 9.6). The dip amount and dip direction of the fold axial plane can be measured in the same way as dipping of beds or faults. Folds are described as *upright*, *steeply inclined*, *moderately inclined* or *gently inclined* depending on the amount of dip of their axial plane (the horizontal axis of the graph in Figure 9.6). Folds are said to be *sub-horizontal*, *gently plunging*, *moderately plunging*, *steeply plunging* or *sub-vertical* depending on the amount of plunge of the fold axis (the vertical axis of the graph in Figure 9.6). Folds are always described using two terms, e.g. *upright* folds with *gently plunging* axes.

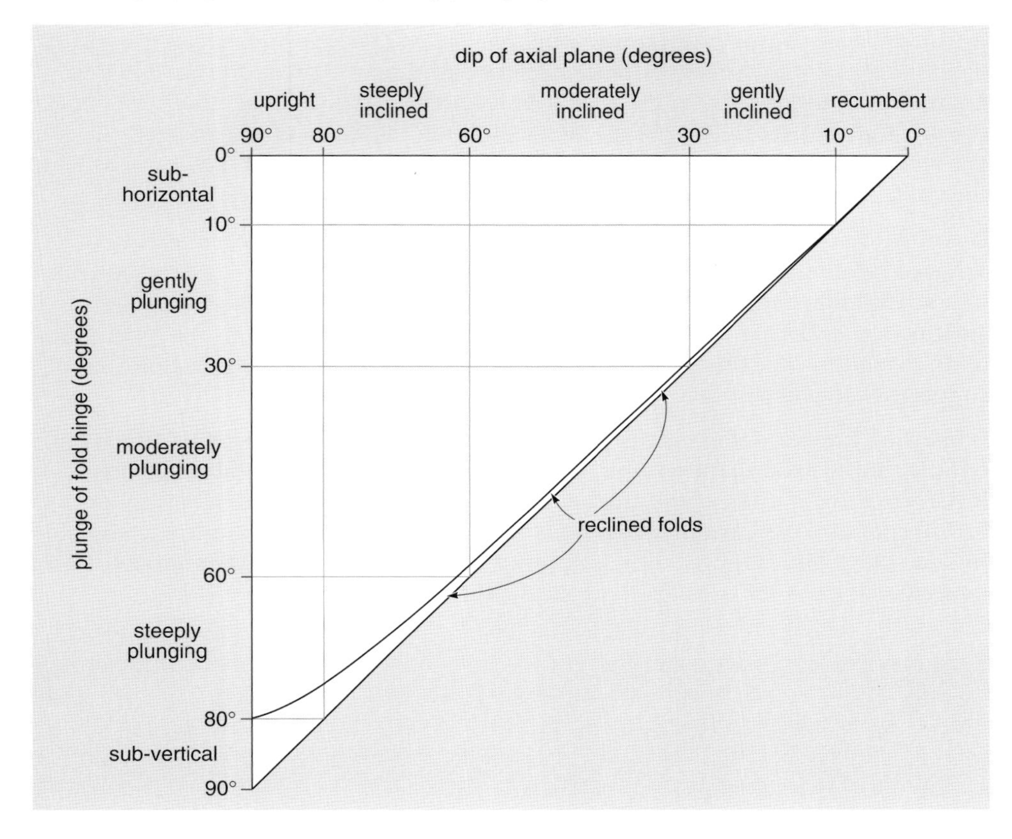

Figure 9.6 A fold classification system. The amount of dip of the axial plane is recorded in degrees on the horizontal axis of the graph; the plunge in degrees on the vertical axis down the left-hand side.

❑ How would you classify the folds in Plate 9.1, using the terms on the graph in Figure 9.6?

■ The fold axial planes are steep, probably greater than 80°, as far as we can estimate from this photograph. We would call these folds upright. (If we found that the axial plane in fact dipped less than 80°, we would then call them steeply inclined.) The fold plunge appears to be low, less than 10°. We call these folds *upright with sub-horizontal plunge*.

There are two specialized but not especially important terms in the graph in Figure 9.6. Folds with almost horizontal (dip $< 10°$) axial planes are called **recumbent folds**. If the amount of plunge of the fold is almost the same as the amount of dip of the axial plane, the folds are said to be **reclined**. Reclined folds therefore plunge directly down the axial plane, parallel or almost parallel to the dip direction. Recumbent folds may or may not be reclined.

Folds must be ductile structures, because beds that are folded are still continuous, with no obvious displacement across any fractures. A word of warning, though. Some folds that you may see in rocks – e.g. those caused by

Figure 9.7 A fine-grained sedimentary rock with a well-developed slaty cleavage. The diagonal tonal banding is the original sedimentary layering, and the fine vertical lines are the slaty cleavage.

the contortions of flowing lava – are not structures that indicate deformation in the structural geological sense. Such folds could be quite common locally. In general though, you would be on reasonably safe ground if you assumed that any folds you see represent deformation in the structural sense.

9.5 Cleavage and schistosity

There is one further very important ductile structure which shows that rocks have been deformed: *rock cleavage*. As we saw in Block 2 Section 8.2, cleavage is a *foliation* found in low-grade metamorphic rocks. The term *schistosity* is used for foliations in higher-grade metamorphic rocks. In slates, it is only possible to see the individual grains with a microscope, whereas in schists it is sometimes possible to see individual grains with the naked eye.

Cleavage is an alignment of platy minerals, typically clays and micas. Very often the alignment of fine-grained platy minerals is almost perfect (Figure 9.7) and the rock is then said to possess a *slaty cleavage*. The term cleavage derives from the fact that slates can be split or 'cleaved' along the grain of the mineral alignment. A general term for both the orientation of the mineral alignment and the splitting direction is the **cleavage plane**.

If you study Figure 9.7, you can see that the slaty cleavage cuts across the original sedimentary bedding. The alignment of platy minerals cannot therefore be a sedimentary feature – it must be a response to some process that has affected the rocks after their deposition. In other words, cleavage is a response to deformation. The absence of significant fractures associated with cleavage suggests that cleavage is a ductile structure.

If bedding has been folded, cleavage planes commonly lie parallel or almost parallel to the axial planes of folds (Figure 9.8a). This is **axial planar cleavage**. If deformation is intense and protracted enough, the same processes that fold bedding planes can also fold cleavage planes. In schists, it is common to see a folded schistosity (Figure 9.8b), frequently accompanied by the formation of new cleavages or schistosities that are axial planar to the folds of the earlier

(a)

Figure 9.8 (a) Folds of bedded rocks in Sardinia (bedding is picked out by the tonal variation) with an associated axial planar cleavage.

(b)

Figure 9.8 (b) Folded schistosity from Bute, Scotland. Here (confusingly!) the tonal variation is not bedding but segregation of metamorphic minerals.

schistosity. In such rocks, bedding can be very difficult to detect because of extensive masking by growth of new metamorphic minerals.

We have described the main types of brittle and ductile structures you are likely to see if you examine outcrops in the field. We now move on to investigate the processes which cause those structures to form.

9.6 Summary of Section 9

- Rock structures provide an important record of events that have affected rocks after their formation.
- Joints are brittle fractures in rocks which show no appreciable lateral displacement across the fracture. They are commonly seen as planes cutting right through the rock mass, and often occur in sets.
- Faults are brittle fractures which show a significant lateral displacement between hangingwall and footwall. They can show dip–slip or strike–slip movement, or a combination of both. The three most important categories of fault are normal faults, reverse faults and strike–slip faults.
- Faults typically occur in arrays, which allow deformation to be regional in extent even though individual faults die out.
- Slickensides and slickenfibres are linear features found on fault planes that enable the orientation of movement to be established, at least in part.
- Shear zones are the ductile equivalents of brittle faults which are often found in rocks that have been deformed at depth.
- Folds are ductile structures that represent permanent deformations of the original sedimentary bedding or other types of layering. Folds are categorized in terms of their wavelength, amplitude, dip of their axial surface, plunge and shape.
- Cleavage and schistosity are alignments of platy (and often tabular) minerals. Both are ductile responses to deformation.
- Cleavage and schistosity nearly always cross-cut bedding, and nearly always have a close relationship with fold geometry.

9.7 Objectives for Section 9

Now you have completed this Section, you should be able to:

9.1 Recognize the main features that show rocks have been deformed.

9.2 Distinguish between joints and faults, folds and cleavage in photographs or in clear field examples.

9.3 Use fault and fold terminology confidently.

9.4 Show that you understand the difference between brittle and ductile structures.

9.5 List the main types of brittle structure and ductile structure that you might expect to see in deformed rocks.

Now try the following questions to test your understanding of Section 9.

Question 9.2 What is the difference between thrusts and reverse faults?

Question 9.3 Describe (*using 20–30 words for each*) the important characteristics of normal faults, thrust faults and strike–slip faults.

Question 9.4 State, giving your reasons, whether joints are brittle or ductile structures. (*Answer in about 30 words*)

Question 9.5 Explain (*in about 40 words*) whether the shear zone in Figure 9.5 shows sinistral or dextral displacement.

10 Rock-deforming processes

We now turn our attention to *process*. How do rocks deform? What controls whether rocks experience brittle or ductile deformation? How do the structures that we see relate to rock-deforming processes? We start by considering two fundamental concepts – stress and strain.

10.1 Stress and strain in rocks

Stress and *strain* are words in common use; we often talk about the 'stresses and strains' of everyday life. However, these terms have precise definitions in physics and rock mechanics. **Stress** is defined as force per unit area and is measured in *pascals* (Pa) – units you have already met for measuring pressure. Stress becomes greater if the force gets bigger, or if the area over which a given force acts gets smaller. As a somewhat dated example, dancers did much more damage to wooden dancefloors by wearing shoes with pointed stiletto heels than through wearing broad flat-heeled shoes. With both types of shoe, the force was the same (determined by the mass of the dancer) but with stilettoes the stress was greater because the area over which the force acted was much smaller. Similarly, it's much easier to push a drawing pin into a wall if you press on the flattened side of the pin!

Pressure is the term used when stresses act equally in all directions. An everyday example is atmospheric pressure, which is the stresses that act equally in all directions on us due to the gravitational attraction between the Earth and its atmosphere. Rocks at any point within the Earth are subjected to a similar confining pressure, which is also gravitational in origin and a function of the mass of rock that surrounds that point. Generally speaking, the deeper you go, the greater the confining pressure becomes. Pressure is one of the major controlling factors in metamorphism, as you saw in Section 8.

When stresses that are *not* equal in all directions act upon a body, they tend to displace it or to deform it, or both. Such stresses generally produce a change in shape. We give the name **strain** to these changes, so stress produces strain. Strain is measured in terms of changes of length, angle, area or volume *per unit* length, angle, area or volume, which means that it is a dimensionless quantity. Strain is simply a change; large amounts of strain mean big changes.

For some materials and under certain circumstances, strain is simply linked to stress. For example, a spring balance stretches when weights are loaded onto it. The amount that the metallic spring stretches (strain) is proportional to the load (stress) – a fact that is actually used to calibrate the spring balance.

Within a spring balance, strain is generally *recoverable*. This means that the material returns to its original shape after removal of the applied stress. The same applies when, say, a rubber band is stretched, as long as the force used is small and is not applied too quickly. Recoverable strain is also known as **elastic deformation**, which is why we sometimes call a rubber band an *elastic* band.

- ❑ Is elastic deformation part of brittle or ductile behaviour?
- ■ Since the rubber band does not develop any fractures, its behaviour is ductile, not brittle.

- ❑ When does a rubber band show non-elastic deformation?
- ■ If you pull the rubber band hard and quickly, it will probably snap. This is non-elastic deformation, because you cannot get it back to its 'unsnapped' state.

- ❑ When the rubber band snaps, is that brittle or ductile deformation?
- ■ Brittle deformation.

As these questions show, elastic deformation is part of ductile deformation. Brittle deformation can never be elastic. Elastic behaviour does happen in rocks. Elastic changes can be generated in laboratory experiments, and rocks also behave elastically when the seismic waves of a distant earthquake pass through them (Section 2.1). Closer to an earthquake centre, however, deformation is more likely to be permanent, both to the rocks and also to the buildings humans inevitably built on them. You even cause a very small elastic strain every time you stand on an exposure!

- ❑ Why don't the structures we see in rocks represent elastic strain?
- ■ Rock structures cannot represent elastic strain, because elastic strain is recoverable and the deformation we see (e.g. folds or faults) is permanent.

After a certain amount of strain, most materials stop behaving elastically and acquire a permanent deformation, and then the link between stress and strain becomes extremely complex (imagine trying to quantify numerically the relationship between stress and strain in the case of the dropped glass). Because of this complexity, the mathematical link to stress through strain is generally closed to geologists. Since we also cannot measure stresses directly in rocks which were deforming millions of years ago, it is very rare that we can establish the exact stress pattern which caused rocks to deform. We know that shape-changes must have been produced by stress, but we usually cannot *quantify* the parameters that link the two.

"Just one more..."

There are three fundamentally different ways in which rocks may show strain (Figure 10.1, overleaf). First, they may change position, simply by moving from one place to another. Movement like this, where every point within the rock moves in an identical way to its neighbour, is known as **translation**. Secondly, a rock body may change orientation, and undergo **rotation** where all points rotate together around some fixed axis. The third type of change is called **distortion**, which is a change of internal shape of the rock. When a rock is distorted, individual points within the rock move relative to one another.

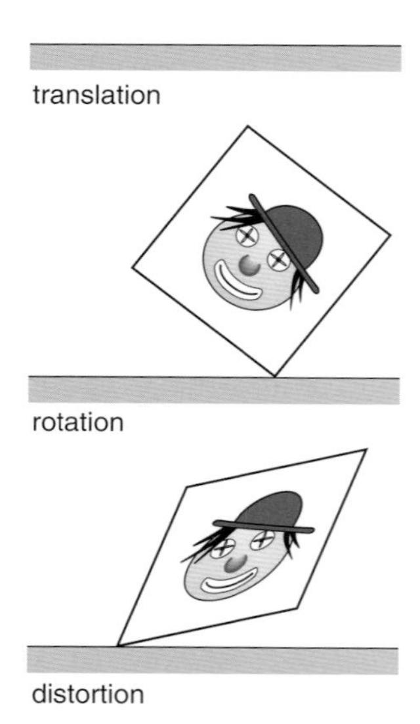

Figure 10.1 The three components of strain: translation, rotation and distortion.

We shall use the general term **deformation** for all these changes taken together, because most strains involve an interplay of translation, rotation and distortion on a range of scales. In the rock record, it is easy to identify distortion, harder to identify rotation, and usually very difficult indeed to identify translation. Be aware that rocks may be deformed even when they have not been distorted – e.g dipping beds record a deformation.

10.2 Strain and the strain ellipse

Distortions are commonly described in terms of the amount of shape change of an originally symmetric shape such as a square or a circle. In Figure 10.2, an undeformed square ABCD containing a circle has been 'squashed' vertically into a rectangle A′B′C′D′. The circle has become an ellipse which, since it records the amount and orientation of strain, is called the **strain ellipse**. The ratio of the long axis to short axis of the strain ellipse, known as its **axial ratio**, gives an indication of the *amount* of strain. The angle the long axis of the ellipse makes with any given reference direction gives the *orientation* of the strain.

❑ Look at the square in Figure 10.2a. How would you describe the changes in length of sides of the square ABCD as it deformed into the rectangle A′B′C′D′ in Figure 10.2b?

■ The top and bottom are longer than they originally were, and the sides have become shorter.

❑ Have any angles changed as the square deformed to a rectangle, either between two adjacent sides (e.g. angle DAB compared with angle D′A′B′) or between the diagonals (e.g. the angle between AC and BD compared with the angle between A′C′ and B′D′)?

■ Adjacent sides were originally at right angles and are still at right angles. However, the angles between diagonals, originally right angles, have changed.

Generally during deformation there are simultaneous changes in both lengths and angles. Area often changes too, but in Figure 10.2b the rectangle has deliberately been drawn so that it has the same area as the square in Figure 10.2a, so this deformation has not involved a change of area.

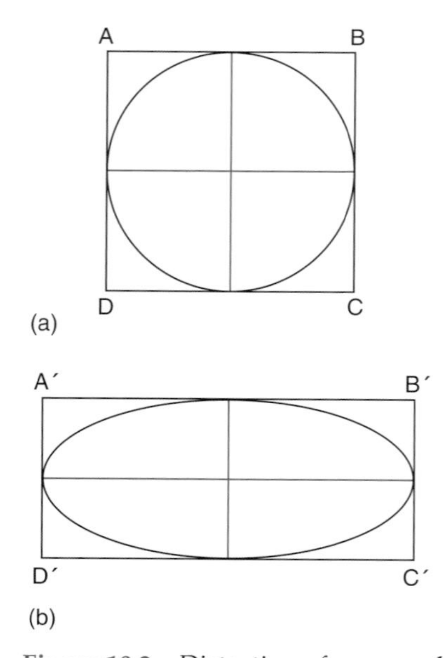

Figure 10.2 Distortion of a square block containing a circle (a) into a rectangular block containing a strain ellipse (b). Note that the areas of the undeformed square ABCD and the deformed rectangle A′B′C′D′ are the same.

10.2.1 Changes in lengths during deformation

Changes in length of lines or linear features in rocks during deformation are recorded by a measure called the **extension**. The extension of any original line is given by the equation:

$$\text{extension, } e = (l - l_0) / l_0 \qquad (10.1)$$

where l_0 is the length of any line before deformation, and l is the length of the same line after deformation.

Numerically, extension is the ratio of the change of length to the original length. Extension is positive if the length of the line increases, and negative if the length of the line decreases. We shall refer to positive extensions as *lengthening* and negative extensions as *shortening*. Extension is commonly expressed as a percentage change in length:

$$\%\ \text{extension} = e \times 100 \qquad (10.2)$$

Question 10.1 What is the % extension of (a) the long axis and (b) the short axis of the ellipse in Figure 10.2b? (*Tip:* compare the length of the diameter of the circle in Figure 10.2a with the ellipse's long and short axes.)

Figure 10.3 Three deformed fossil belemnites. The darker fragments are part of the shells of the fossils, the white areas are where calcite crystals have grown in between the fragments.

There is a practical difficulty in calculating extensions in deformed rocks. We can usually measure the length of some particular feature in the deformed rock, but generally we cannot be exactly sure how long that feature was *before* deformation. However, in many instances we can get an estimate. For example, Figure 10.3 shows three stretched fossils from deformed limestones in Switzerland. We do not know exactly how long each of these tube-like shelly fossils (called *belemnites*) was before strain, but we can measure the deformed length. We can also make a close estimate of the original length of the shells by adding together the lengths of each of the dark fragments. Notice that the amount of extension of each belemnite is different, and depends on its orientation. Many linear features in rocks, such as fossils, mudcracks, or a stretched igneous dyke, may allow rough estimates of extension to be made.

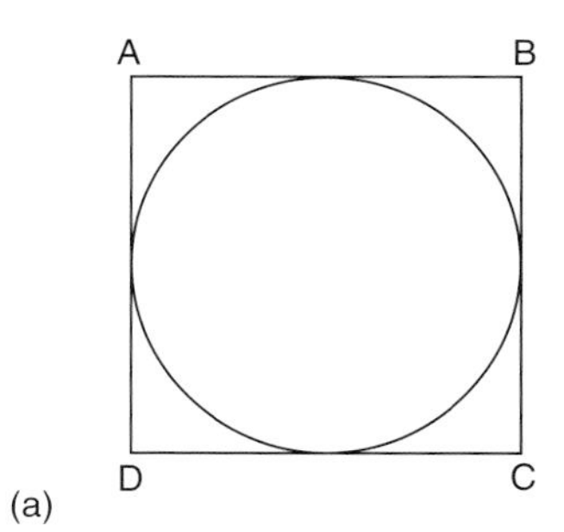

10.2.2 CHANGES IN ANGLES DURING DEFORMATION

Angular changes are quantified by reference to two lines that were originally perpendicular in the undeformed rock.

Figure 10.4 shows an original square containing a circle deformed into a rhomb containing a strain ellipse. Deformation has taken place by the process of *shearing* (similar to pushing over a pack of cards). The amount of angular change, the angle of **shear**, ψ (the Greek letter *psi*, pronounced 'sigh'), can be calculated from any deformed right angle. For example, we could use the angle the side of the rhomb makes with its base because those lines were originally perpendicular before deformation. The value of **shear strain**, γ (the Greek letter *gamma*) is given by:

$$\gamma = \tan \psi \qquad (10.3)$$

Question 10.2 In Figure 10.4b, what is (a) the amount of shear strain, γ, of the deformed square, (b) the percentage extension of the long axis of the resultant strain ellipse, and (c) the percentage extension of its short axis? (*Tip:* as in Question 10.1, calculate extensions by comparing pre-deformation with post-deformation lengths.)

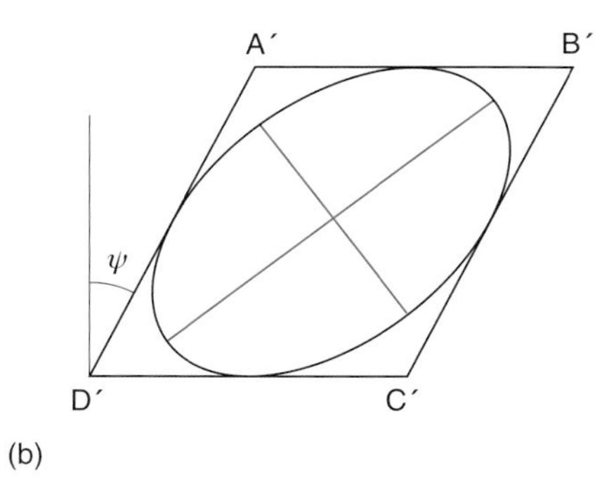

Figure 10.4 Distortion by shearing of a square block containing a circle (a) into a rhomb containing a strain ellipse (b). Note that the area of the undeformed square ABCD and the deformed rhomb A´B´C´D´ are the same.

For the same reasons as extension, shear is also hard to quantify in real rocks. However, shear strain can be calculated wherever objects of known shape – particularly those containing natural right angles – have been deformed, like the fossil *trilobites* in Figure 10.5. Like extension, the amount of shear is generally different in different orientations.

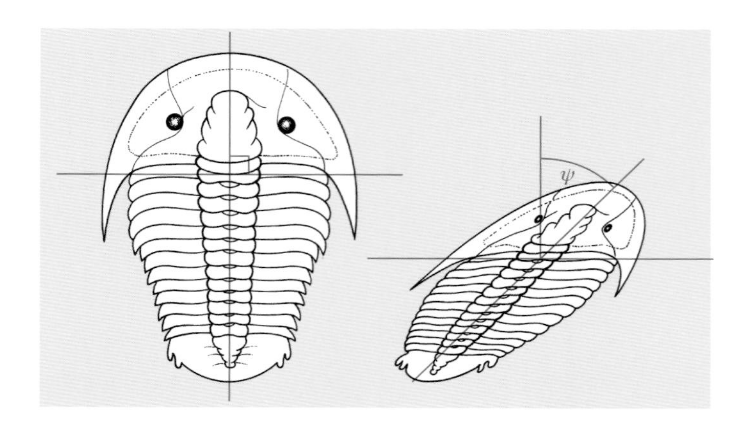

Figure 10.5 Undeformed and deformed fossil trilobites of the same species.

One further point is very important. The ellipse formed by shear in Figure 10.4 is a similar elliptical *shape* to the ellipse shown in Figure 10.2; the main difference is its *orientation*. If we found an elliptical shape in real deformed rocks (without the accompanying deformed square), we would be unable to tell if it had been formed from a circle by squashing, or by shearing followed by rotation (or for that matter by squashing followed by rotation). This highlights the main practical difficulty in linking strain to strain-causing processes; finding a deformed shape in rocks simply does not give us enough information to identify the process that produced the shape.

10.2.3 Changes in area during deformation

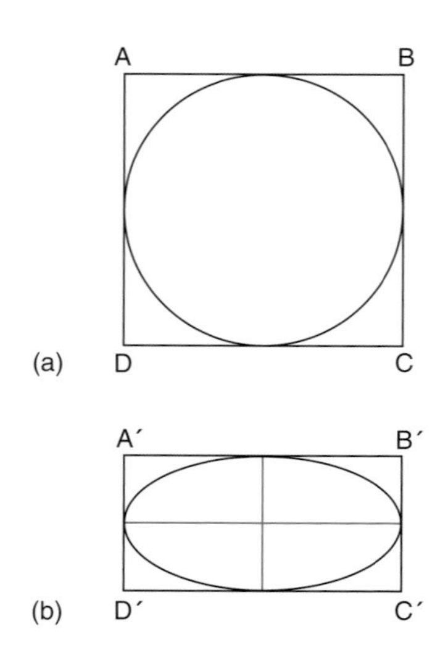

Figure 10.6 Distortion of a square block containing a circle (a) into a rectangle containing a strain ellipse (b) by area reduction. Note that the area of the undeformed square ABCD and the deformed rectangle A′B′C′D′ are *not* the same.

The two deformations we have so far considered have not actually involved a change in area, but you can appreciate that such a change could happen, and is likely in nature. In Figure 10.6, the short axis of the ellipse has shortened, but the ellipse long axis is the same length as the diameter of the original circle. The area of this ellipse must therefore be less than the area of the original circle.

Question 10.3 What is the axial ratio of the strain ellipse in Figure 10.6b?

Area change adds yet another complication to our strain analysis. In general, we can't tell what part, if any, area change has played in deformation simply by measuring deformed shapes. The ellipse in Figure 10.6b is very similar in shape to both the ellipse in Figure 10.2b and to that in Figure 10.4b, yet each of the three ellipses formed by a different process. You can no doubt see that if the amount of deformation in each process was just right, the three strain ellipses would have *exactly* the same shape.

This Section might seem to you to be somewhat negative in tone – why bother trying to measure strain at all if there are so many reasons why the measurements will not lead us to understand strain-causing processes? The answer is empirical; even though strain measurements may not yield precise, quantitative information about processes, they do provide very useful data with which to investigate processes.

10.3 Progressive deformation

So far we have considered deformation as a single event, which we have seen could happen through squashing, by shear or through volume change; real rock deformation is more likely through a combination of all three. Yet we should also consider another component: *time*. In nature, rocks deform by the superimposition of successive increments of strain, rather than in just one single event.

We must think of the transition from circle to strain ellipse as the superimposition of a series of small strains one upon another. We refer to deformation generated through successive increments of shape change as **progressive deformation**. Each increment can be represented by an almost-circular strain ellipse. These ellipses may be superimposed *coaxially*, where the ellipse axes always lie in the same orientation (Figure 10.7a). We call this **irrotational strain**. Alternatively, the strain ellipses may be superimposed *non-coaxially*, where successive ellipse axes lie in different orientations (Figure 10.7b). This is **rotational strain**. Three-dimensional, progressive, rotational strain is the usual case in naturally deforming rocks.

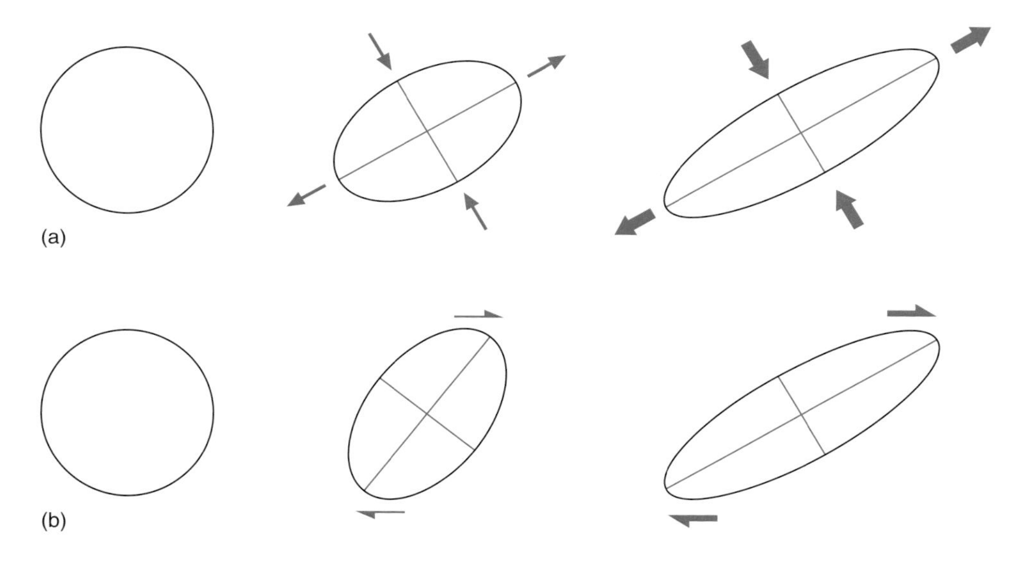

Figure 10.7 Cartoons showing progressive irrotational and rotational strain. In (a), small strain increments have been superimposed coaxially, whilst in (b), successive increments of shear have caused small strain increments to be superimposed non-coaxially. Note that in terms of shape and orientation, the two final ellipses in (a) and (b) are identical, even though they formed in different ways.

10.3.1 Investigating strain

During progressive deformation, lines in some orientations get longer whilst lines in other orientations get shorter. In some orientations, lines start off shortening and end up lengthening. Some lines never change length. The best way really to appreciate these changes is to investigate them for yourself.

Activity 10.1

Try Activity 10.1 to explore some special features of extension in strain ellipses.

Activity 10.2

Now tackle Activity 10.2, to investigate length and angle changes during progressive shearing.

10.4 THE MAIN FEATURES OF STRAIN

Activities 10.1 and 10.2 should have convinced you that how a line in a rock (or plane, in three dimensions) deforms depends not only on what process caused the deformation but also on how that line (or plane) was oriented in the rock prior to deformation. During even a relatively straightforward process like shearing rocks with uniform properties, lines in different orientations extend by different amounts. Angles between lines in some orientations change rapidly whilst angles between others change slowly.

When rocks deform naturally, they rarely do so in a straightforward, tidy manner. Usually, irrotational and rotational strain increments are both involved, possibly at the same time. Commonly, the rock will lose or gain volume during deformation as well. You can no doubt see that it is unwise to think in terms of say 'structures associated with compression', because a process-based term like this takes no account of the complex way in which natural materials change shape. All sorts of changes are 'associated with compression', including shortening of beds in some orientations, lengthening in others, as well as reverse, strike–slip and even normal faulting.

We end our discussion on strain with several important conclusions which you really must bear in mind whilst studying deformed rocks, and in particular as you study the next two Sections:

- Whenever rock shortens in one direction, it *must at the same time* lengthen in at least one other direction, if volume is to be conserved. The converse is also true: lengthenings demand shortenings.
- Rotated features, such as dipping beds, don't automatically imply rotational strain. Lines and planes in most orientations rotate during irrotational strain as well.
- In general, most natural distortions are formed by a combination of non-coaxial strain (i.e. rotational strain) and volume change.
- Most natural strains can be considered as three-dimensional combinations of distortion, translation, and rotation.
- Acute angles and strongly stretched linear features generally imply more strain. For example, tight folds show greater strain than open folds.

10.5 SUMMARY OF SECTION 10

- Stress is defined as force per unit area, and is measured in pascals. Stresses acting equally in all directions are known as pressure. The deeper you go within the Earth, the greater the confining pressure.
- When stresses that are not equal in all directions act upon a body, they tend to deform it.
- Strain is produced by stress. Strain is measured in terms of changes of length, angle, area or volume per unit length, angle, area or volume, and is a dimensionless quantity.
- Elastic strain is a recoverable form of ductile behaviour. Most rock structures are non-reversible, and therefore not elastic.
- There is no simple mathematical link between stress and strain for real deformations.
- Rocks may show strain as translation, rotation or distortion. Most deformations are a combination of all three.
- During deformation, there are generally simultaneous changes in both lengths and angles. Changes in length are called extensions; changes in angle are called shears. A circle will deform into an ellipse – the strain ellipse – which reflects the amount and orientation of strain.

- Progressive deformation is generated through successive increments of shape change. Small strain increments might be superimposed coaxially as irrotational strain, or non-coaxially as rotational strain. Progressive rotational strain is the usual case in naturally deforming rocks.
- Whenever rock shortens in one direction, it *must* lengthen in another direction at the same time, if volume is to be preserved.
- Rotated lines or planes do not automatically imply shearing. Most lines and planes also rotate during irrotational progressive deformation.
- Acute angles and strongly stretched linear features generally imply greater strain.

10.6 OBJECTIVES FOR SECTION 10

Now you have completed this Section, you should be able to:

10.1 Distinguish between the concepts of stress and strain, and understand the difference between recoverable and permanent strains.

10.2 Explain the relationships between strain ellipses, extension and shear.

10.3 Use lengthened or shortened linear features in rocks to determine extension, and changed right angles to determine shear.

10.4 Explain how natural strains can be considered as amalgamations of three-dimensional strain increments.

Now try the following questions to test your understanding of Section 10.

Question 10.4 Figure 10.5 shows two *trilobite* specimens, one deformed and the other undeformed. Determine the shear strain, γ. Why can't you measure the extension of the deformed trilobite?

Question 10.5 Use Plate 10.1 to identify translation, rotation and distortion in real rocks. Label features which show each of the above structural components. (*One* feature only per component.)

11 HOW STRUCTURES FORM

How rocks respond to stress depends on many physical variables, including:

- the *temperature* and *pressure* at which the rock deformed;
- the *rate* at which the rock is deformed;
- the *properties of the rock material* itself.

The physical properties of deforming rocks also change as the rocks deform. This might be because new minerals grow, or because beds change thickness, or in particular because the nature of the fluids occupying the spaces within the rock changes. There are many possible variables – too many to consider in detail here. We shall concentrate on two broad categories: 'external properties' imposed on rocks from outside, e.g. temperature; and the 'internal properties' of the rock body itself, in particular its composition.

Activity 11.1

Get a feel for the way strain is influenced by the physical properties of everyday materials by trying Activity 11.1.

11.1 Strain and the effects of pressure, temperature and time

The confining *pressure* at which a body of rock deforms depends on how deeply it is buried during deformation. Generally speaking, the greater the depth, the greater the mass of the rock overburden and the higher the confining pressure. The *temperature* at which rocks deform depends both on the depth of burial *and* on the geothermal gradient. Hot rocks generally behave in a ductile way, cool rocks in a brittle way (you saw a similar effect in chocolate in Activity 11.1). Rocks deformed at low temperatures and pressures at or near the surface of the Earth show far more evidence of brittle failure than those deformed at depth.

❑ What structures should be common in rocks that have deformed near the Earth's surface?

■ Joints and faults, since they are the common brittle structures.

❑ Why would we expect major faults to pass into shear zones at depth?

■ The higher pressures and temperatures that exist at depth favour ductile rather than brittle deformation, and shear zones can be considered as ductile equivalents of faults.

The *rate* at which rocks deform is called the **strain rate**. The amount of change over a given period of time is measured as *strain per second*. Since strain is dimensionless, strain rate has units of s^{-1} ('per second').

Rock sequences naturally deform over millions of years, not seconds. For example, the Alps represent a tectonic event in which the crust shortened to about half its original length and approximately doubled its thickness in 10–20 Ma. So strain rates are always very small numbers.

❑ What range of strain rates does the Alps event represent?

■ Assume the Alps record a doubling in thickness, from an original thickness t to a new thickness of $2t$. This is an extension of $(2t - t)/t = t/t = 1$, in 10–20 Ma. The strain rate is therefore between 1 per 10 Ma and 1 per 20 Ma. There are about 3×10^{13} seconds in 1 Ma, so in conventional units (s^{-1}), the strain rate ranges between $1/10(3 \times 10^{13})\ s^{-1}$, which is approximately $0.3 \times 10^{-14}\ s^{-1}$, and $1/20(3 \times 10^{13})\ s^{-1}$, or approximately $0.17 \times 10^{-14}\ s^{-1}$.

Such strain rates, roughly $10^{-14}\ s^{-1}$ to $10^{-15}\ s^{-1}$, are typical for naturally deformed rocks. $10^{-16}\ s^{-1}$ is slow, whilst $10^{-13}\ s^{-1}$ is fast. Under certain circumstances, natural strain rates can be much faster. For example, recent OU research has shown that magma movements are causing the summit region of Etna to deform at a staggeringly fast rate of $10^{-10}\ s^{-1}$.

Question 11.1 How many *orders of magnitude* different would the strain rate be if the Alps had undergone twice as much shortening and vertical thickening as they actually have?

Generally speaking, rocks deformed at high natural strain rates show more brittle behaviour than those deformed at low strain rates. The film in Activity 11.1, and some brands of toffee, show this effect. If you hit toffee with a hammer it will break, but you can also bend a toffee bar as long as you do it slowly.

External controls act in combination. Rocks deformed quickly at shallow depths usually deform in a brittle way, whereas rocks deformed slowly at depth will usually deform in a ductile way. If you warm the toffee and flex it slowly, it should always bend, never break.

11.2 STRAIN AND ROCK COMPOSITION

If you spent some time looking at deformed rocks in the field, you would surely see a wide variation in structures – folds of varying wavelengths, amplitudes and shapes, some beds cleaved whilst others are not, faults here but not there, and so on. In any compact geographical region, it is almost inconceivable that the rocks have experienced widely different temperatures, pressures and strain rates during deformation. Variations in structures on a bed-by-bed basis cannot wholly be explained by changes in external physical controls. For an explanation, we must look at the physical properties of the rocks themselves.

Rock *composition* has a profound effect on deformation. Some rock types are almost always rigid and tend to resist deformation; others almost always flow during deformation. This attribute is termed **competence**. Competent lithologies are stiffer, flow less easily and break more readily than incompetent lithologies. Sandstones, limestones and most igneous rocks tend to deform in a competent way during deformation, whilst mudstones, shales and evaporites generally behave in an incompetent way.

❑ In a deforming succession of interbedded sandstones and shales, would you expect faults to initiate in the sandstones or in the shales? Why?

■ In the sandstones. This is because given similar pressure, temperature and strain rate, faulting (brittle failure) is more likely to occur first in competent sandstones rather than incompetent shales.

In successions of sedimentary rocks, every fold and every fault we look at is similar yet not identical to its neighbour. The clue to how rocks deform lies in this word *similar*. Structures are controlled by subtle variations in the rock, such as different thicknesses from bed to bed, changes in thickness of individual beds, particularly the thicker ones, or changes in spacing between beds. The presence and composition of pore fluids also have a marked effect on how rocks deform. The successions of rocks in different places are never absolutely identical, so the outcome of deforming rock successions can never be the same. Furthermore, the physical features we have discussed so far are only *some* of the factors that determine what type of structures develop.

An obvious way forward from here would be to look at the simplest or most uniform of rock lithologies, and examine how variations in both physical parameters and strain components influence which structures form, then extend our investigation by degrees to more and more complex interlayers of rocks. However, we would quickly find that way to be lengthy, extremely complicated and generally unsatisfactory. Instead, we shall look at the structures discussed in Section 9, and investigate the processes that lead to their particular formation. We will only get an insight into rock deformation and not an exhaustive account, but it will illustrate effectively how structures form.

11.3 HOW BRITTLE STRUCTURES FORM

11.3.1 JOINTS

Joints occur in every rock type that is found on the surface of the Earth, so they are clearly not just dependent on lithology. As brittle structures they belong to the upper, colder parts of the crust – we have no evidence for the existence of joints deep down. Where their origin is clearest, they seem to be associated with volume changes within the rock. This is evident, for example, from the polygonal columnar joints seen in some lava flows (Section 6.1), which contracted as they cooled, or in the hexagonal cracks seen on the top of dried mud at the roadside.

Since joints are so common, and since we know that the rocks we see exposed at the surface have usually been buried to some degree, many structural geologists believe there is an association between burial, subsequent uplift and erosion (i.e. exhumation), and joint formation.

11.3.2 Boudins and competence contrast

Materials with a completely uniform composition, such as Plasticine, deform in the simplest way. What would happen if we deformed a block of Plasticine? Of course we would not expect joints, faults or even folds to develop – the Plasticine block would simply shorten in some directions and lengthen in others.

Sedimentary rocks, however, consist of different beds with different compositions and thicknesses, and therefore different competences. When sedimentary rocks are strained, the effects of **competence contrast** can be seen. Suppose we stretch a layered composite of materials of different competences parallel to their length (Figure 11.1). The competent layers may fracture, and separate into lozenge-shaped pieces called **boudins** (pronounced '*boo*-dans'), from the highly descriptive French term for a type of sausage.

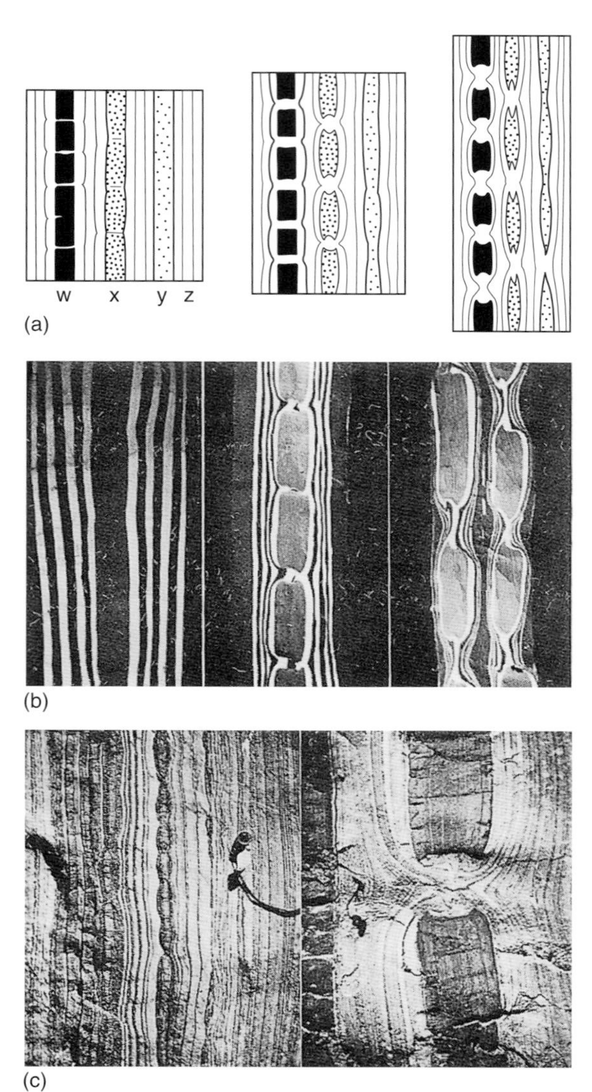

Figure 11.1 Development of boudins. (a) Diagrammatic cross-sections through boudins in materials with a competence contrast (competence of materials: w>x>y>z; z is the matrix). (b) Three stages through a laboratory experiment with layers of different competence in plastic modelling materials. (c) A real example from metamorphosed limestones in Namibia.

Where the competence contrast is high, as it is between layers (w) and (z) in Figure 11.1a, rectangular boudins separated by fractures form in the most competent layer. Where the competence contrast is low, as it is between layers (y) and (z) in Figure 11.1a, the more competent layer flows and necks, and can even stretch without breaking. The theoretical situation shown in Figure 11.1a can be duplicated in laboratory experiments (Figure 11.1b), and is also seen in real rocks (Figure 11.1c).

11.3.3 STRESS AND FRACTURE ORIENTATION

The boudins in Figure 11.1, like the joints in Figure 9.1, show fractures that formed almost at right angles to the bedding planes. Do faults also form at right angles to bedding, like joints and boudins?

To answer this question, we must delve into stress in rather more detail. Theoretically, the stresses acting on a volume of rock can be resolved into three components, the three *principal stresses*, which always act at right angles to one another. These principal stresses are labelled σ_1, σ_2 and σ_3 (the Greek letter is 'sigma'). Whilst the orientation of each principal stress is established – each is mutually perpendicular to the other two – the magnitudes of the stresses are not. Each one of σ_1, σ_2 and σ_3 might be either *compressive* or *tensile*, that is each could act to shorten or lengthen the rock *in that orientation.*

In the special case where there are no tectonic stresses, the only stress acting at any point is the pressure of the rock overburden. By definition, pressure means stresses acting equally in all directions (Section 10.1), so $\sigma_1 = \sigma_2 = \sigma_3$. All three principal stresses will be compressive – they are acting to try to shorten the rock and reduce its volume.

More generally, and certainly whenever rock is deforming tectonically, the magnitude of the three principal stress components will *not* be equal. One of the three has to be the most compressive (which only means more compressive than both the other two) and another has to be the least compressive (less compressive than the other two). The third will have an intermediate value between these two extremes. By convention, we label unequal stresses such that σ_1, σ_2 and σ_3 show the greatest, intermediate and least *compressive stress* respectively (Figure 11.2a).

In real situations, it is often the case that σ_3 is tensile whilst σ_1 is compressional, but note that we don't actually know whether σ_3 (or σ_2 and σ_1 for that matter) is actually tensile or compressive, simply that both σ_3 and σ_2 are, by definition, *less compressive* (and thus more tensile) than σ_1.

What happens when real rocks are being deformed, if σ_1, σ_2 and σ_3 are all compressive and if σ_1 is very much more compressive than σ_3? At first the rock shortens elastically parallel to σ_1 and lengthens elastically parallel to σ_3, until it may fracture. We know from theory and experiments that under compressive stresses a fracture will *always* develop at an acute angle to σ_1 (angle α in Figure 11.2b). The angle α is always less than 45°, typically in the range 20°– 40°. The fracture plane strikes parallel to σ_2. Although only one orientation of fracture plane is shown in Figure 11.2b, there is a mirror-image fracture which is just as likely to form. (Imagine viewing the page from behind, or hold a mirror up in front of the page to see how it looks.) Often fractures form in both possible orientations *at the same time*. When two such fractures form, they meet in an X-like pattern on surfaces normal to σ_2. We then describe them as *conjugate* fractures, as long as they formed at the same time.

The key to understanding the formation of faults lies in recognizing that the presence of layers within deforming rocks actually controls the *orientation* of the stress field. The layers might typically be beds in sedimentary rocks, or perhaps

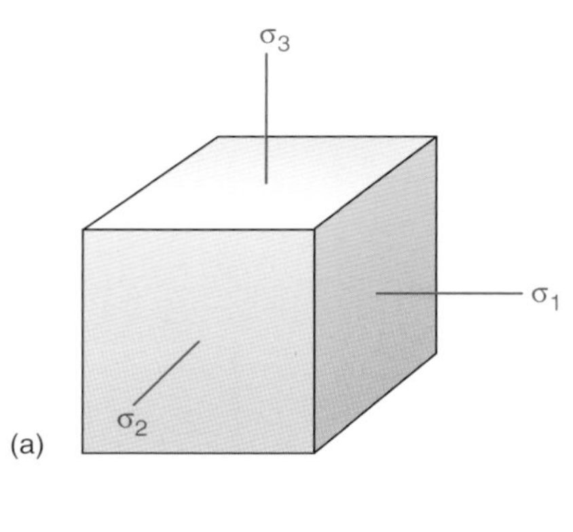

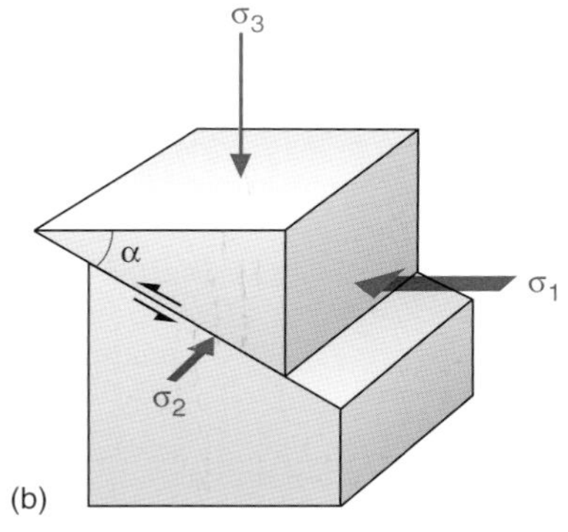

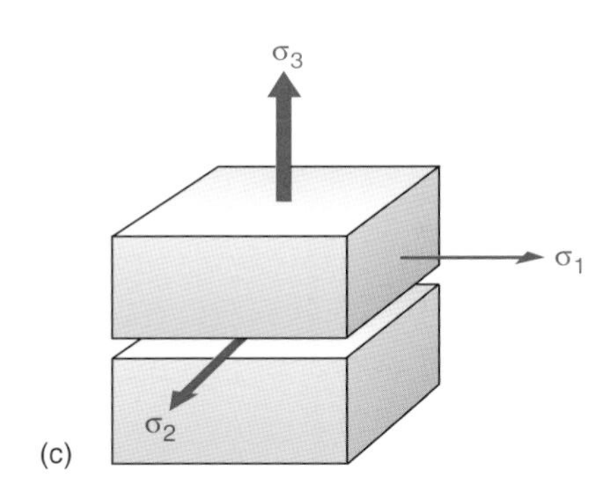

Figure 11.2 The development of fractures can be related to the magnitude of the principal stresses. (a) The three principal stresses σ_1, σ_2 and σ_3 are always mutually perpendicular, but each may be either tensile or compressive. However, σ_3 is always the most tensile and σ_1 always the most compressive stress. (b) Under experimental conditions, failure under a compressive stress field (all principal stresses compressive, $\sigma_3 < \sigma_2 < \sigma_1$) causes a fracture plane to develop which lies at an acute angle α to σ_1. α is typically 20°–40°. (c) Failure under tensile stress field (all principal stresses tensile, $\sigma_3 > \sigma_2 > \sigma_1$) produces a fracture plane normal to σ_3.

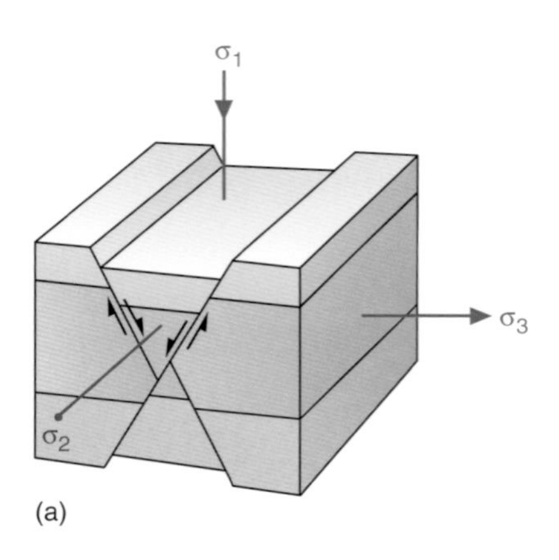

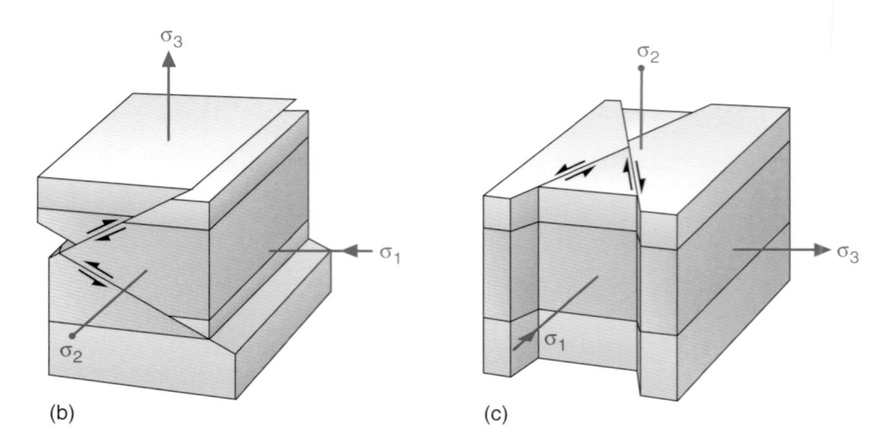

Figure 11.3 Three main types of fault that can initiate in layered rocks in response to different orientations of principal stress. (a) Conjugate normal faults form when the maximum principal compressive stress σ_1 is normal to layering. (b) Conjugate thrusts form when σ_1 lies along layering and σ_3 is normal to layering. (c) Conjugate strike–slip faults form when both σ_1 and σ_3 lie parallel to layering.

the surface of an igneous pluton, or a metamorphic schistosity, and so on. Usually one of σ_1, σ_2 and σ_3 is constrained to act at right angles to the layering, whilst the other two act parallel to it.

Let us consider a case where sedimentary rocks near the Earth's surface are just starting to deform, and beds are still almost horizontal. Let us also assume that in this case σ_1 is the principal stress which is oriented at right angles to bedding (so σ_2 and σ_3 must both lie parallel to bedding).

❑ In this case, is the rock body trying to lengthen or shorten along its layers?

■ Since σ_1, the maximum compressive stress, is normal to layering, the stresses acting along the layering (σ_2 and σ_3) must be comparatively tensile and acting to lengthen the layers.

❑ If beds are trying to lengthen, and deformation is brittle, what sort of structure would you expect to develop?

■ Normal faults should develop.

Since failure planes always make a small angle with σ_1, normal faults should initiate as planes dipping steeply through the sedimentary layers (Figure 11.3a). One or both of the conjugate fractures may develop – if conjugate normal faults develop, a downthrown block or graben will form.

Let us now imagine a situation where σ_3 is normal to bedding. Here, we would expect beds to shorten horizontally and thicken vertically. If brittle deformation occurs, we would expect reverse faults to develop. Since failure planes always make a small angle with σ_1, reverse faults should initiate as planes dipping gently through the sedimentary layers. Thrust faults – low-angle reverse faults – should therefore form (Figure 11.3b). These theoretical predictions fit very well with observed normal and reverse fault orientations, especially in deformed sedimentary rocks.

To complete the analysis, conjugate strike–slip faults should initiate when the principal stresses lie in a third specified orientation (Figure 11.3c).

11.3.4 Joint orientation revisited

If the above analysis is sound, why then do we so often see joints which form *at right angles* to bedding in sedimentary rocks, as in Figure 9.1? Why don't joints always form at an angle to bedding, in the same orientation as faults?

The first answer is – sometimes they do! Many rocks, particularly folded rocks, carry conjugate joint sets that lie in the 'correct' orientation for the fold to have formed by shortening. However, this is clearly not the explanation for every joint set. The beds in Figure 9.1 are not obviously folded, and the joints clearly lie at right angles rather than oblique to bedding.

The second answer is rather complicated. We know from experiments that if *all three* principal stresses are actually tensile, rather than compressive, rocks fail in the manner shown in Figure 11.2c. Fractures form perpendicular to the greatest tensile stress (the least compressive stress, σ_3), opening up a fracture in the rock known as an **extension fracture**. Bedding-normal joints therefore suggest failure under conditions where the principal stresses were all tensile, and where σ_3 lay parallel to bedding.

How can this come about, when tectonic stresses and even the stresses due to rock overburden are compressive? It happens because of the extremely complicated relationship between stress and strain in real situations. Stresses are actually set up in rocks *as a consequence* of their having been deformed. These stresses are recoverable over a period of time. For initially compressed rock, such recoverable stresses will be tensile in character. If a rock is buried and/or deformed and then brought back towards the surface, a situation occurs where the magnitudes of the 'external' stresses acting on the rock are matched and then exceeded by the stored recoverable stresses within the rock body. When these 'internal' tensile stresses sufficiently exceed the 'external' compressive stress, extension fractures will result.

The previous paragraph is heavy going and delves into topics that are really beyond the scope of this Course. The main point you should take from it is one we made earlier – that joints at right angles to bedding can develop as a result of burial followed by uplift, with its associated release of elastically stored stresses.

11.3.5 Fault profiles

Faults are rarely as straight and planar as they appear on maps, because of the highly variable physical properties of rock layers. In particular, they typically show variations in dip. Figure 11.4 shows how faults often initiate and develop, using thrusts as an example.

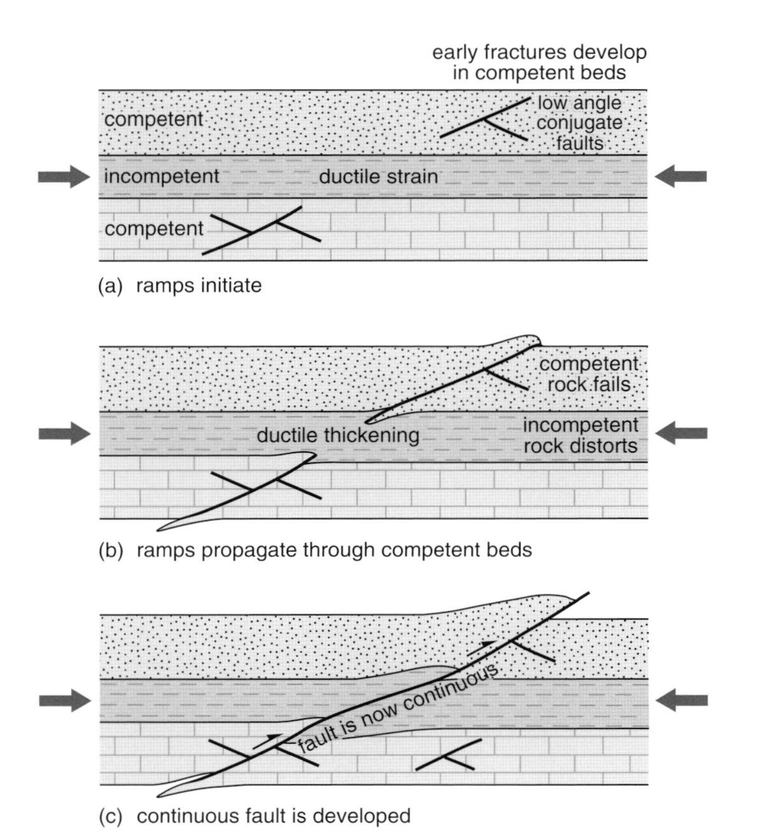

Figure 11.4 Cartoons illustrating stages (a), (b) and (c) in the initiation and development of a thrust fault in mixed sedimentary layers. The red arrows indicate that shortening is parallel to the layering. A conventional system of rock ornaments has been adopted; stipple indicates sandstone, dashes indicate shale, and a brick-like ornament indicates limestone.

Here a succession of sub-horizontal, undeformed sediments is shortened parallel to its layers. Fractures initiate in competent beds of limestone and sandstone, at acute angles to bedding in accordance with the theory discussed above (Figure 11.4a). Either one of a pair of conjugate fractures could develop; in this example, leftward-dipping fractures grow. The incompetent shale that lies between competent beds must also be shortening at the same time, most probably by volume loss through expulsion of pore fluids from between the grains. As shortening continues, the newly formed fractures propagate right through the competent beds (Figure 11.4b), and eventually join along failure planes in the less competent shale (Figure 11.4c). The thrust becomes continuous, allowing the hangingwall to move over the footwall.

The characteristic profile of a thrust is therefore a stepped one consisting of **flats** and **ramps**. Flats are those sections of the thrust plane that lie parallel or almost parallel to beds, whilst ramps are those sections that cut up through beds (Figure 11.5). Invariably, flats occur in incompetent horizons and ramps in more competent horizons. Most thrusts ramp upwards through their footwalls in the direction of transport of their hangingwalls. Note that flats and ramps are defined relative to orientations of beds, not necessarily to the horizontal.

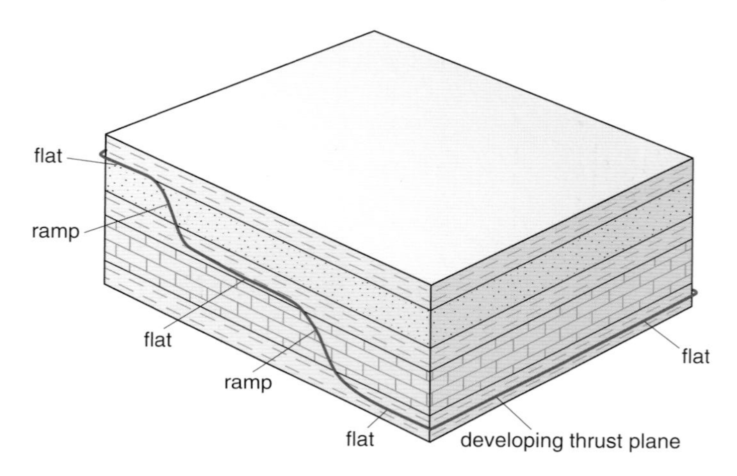

Figure 11.5 Flats and ramps develop in beds of contrasting competences, leading to a stepped fault profile.

Activity 11.2

Now do Activity 11.2, where you view the middle part of the video sequence *Structural geology without tears* on DVD 2.

11.4 How folds form

Folds form in two fundamentally different ways: as a mechanical consequence of fault displacement and as a mechanical consequence of shortening along layers.

11.4.1 Fault-bend folds

Activity 11.3

Surprising things happen when the hangingwall of a stepped thrust moves over its footwall. Discover them now by doing Activity 11.3.

The open, upright anticline you constructed in Activity 11.3 is a direct consequence of fault hangingwall movement. Both anticlines and synclines can form in this setting. This example was for movement up a stepped thrust, but as you could imagine, and can readily verify if you have the time, folds can form by hangingwall movement down a stepped normal fault. Both are examples of **fault-bend folds**.

❑ Look at the folds you constructed in Activity 11.3. How far down through the whole rock succession do these folds occur?

■ The folds are restricted to the hangingwall block only. The footwall block is not folded.

11.4.2 Tip-related folds

Now look again at Plate 10.1, which shows a thrust-related fold from Broad Haven in south-west Wales. First, you must make sure you can recognize the fault plane itself. This is an approximately horizontal line which passes through the centre of the outcrop, below the anticlinally folded sandstone beds. In the right-centre of the photograph, it marks a break between almost horizontal beds above and steeply-dipping beds below. Some prominent beds are displaced some 2–3 m to the left along this fault, so the hangingwall transport direction is from right to left (actually northwards).

❑ Is this thrust folded?

■ No.

❑ Are beds in the footwall of this thrust folded?

■ Yes.

Clearly, some fold-forming process other than simple fault-bend folding is needed to explain why the footwall as well as the hangingwall has been deformed. This mechanism is illustrated in step form in Figure 11.6, which you should recognize from the video sequence in Activity 11.2.

Immediately ahead of the tip of an advancing thrust lies an area which so far has not been deformed. As the 'envelope' of deformation approaches, strain increases in this area from zero through low strains towards intermediate strains. Strain rate also increases, and we should expect initial ductile strain (at low strain rates) to give way to brittle strain (at higher strain rates). Deformation will initially be achieved by ductile processes – folding and volume change rather than faulting. Figure 11.6a, b shows a fold developing ahead of the advancing thrust tip.

❑ Is this fold pair (anticline and syncline) symmetric or asymmetric?

■ Asymmetric. The common, leftward-dipping limb is steeper than the rightward-dipping limbs (and eventually becomes overturned).

❑ How does the fold asymmetry relate to the direction of thrust displacement?

■ The geometry of the asymmetric anticline and syncline is consistent with the nature of the thrust displacement – top towards the direction of thrust advance.

As brittle failure is reached (Figure 11.6c), the thrust propagates forward through the newly formed fold. The upper, anticlinal part of the fold pair becomes part of the thrust hangingwall, and is displaced away from its syncline which is left behind in the footwall (Figure 11.6d).

Many folds initiate in front of thrust tip-lines. They are typically asymmetrical, overturned in the direction of the thrust transport. Most folds that form during thrusting are a combination of two processes – fault-bend folding and tip-related folding.

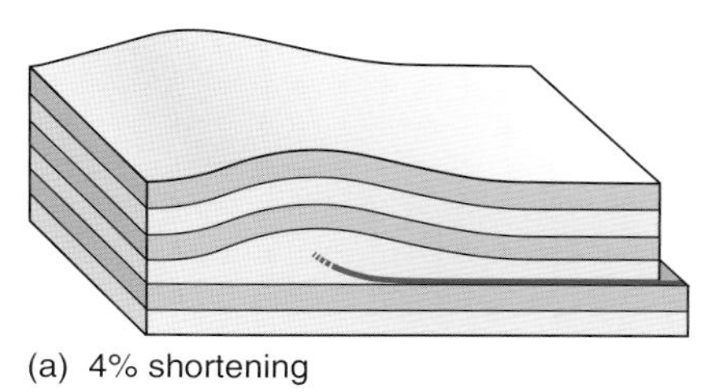

(a) 4% shortening

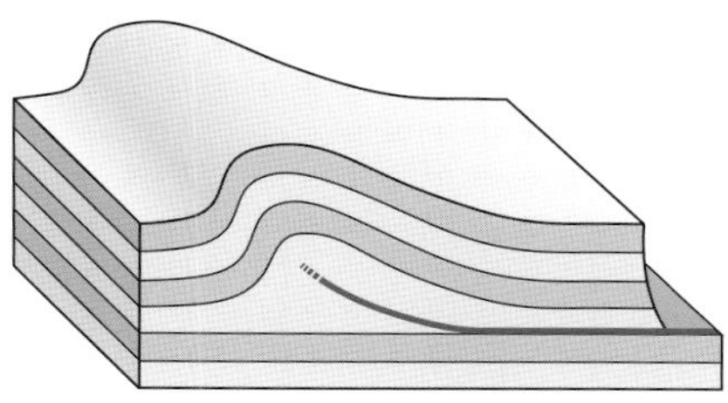

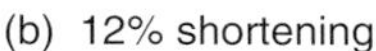

(b) 12% shortening

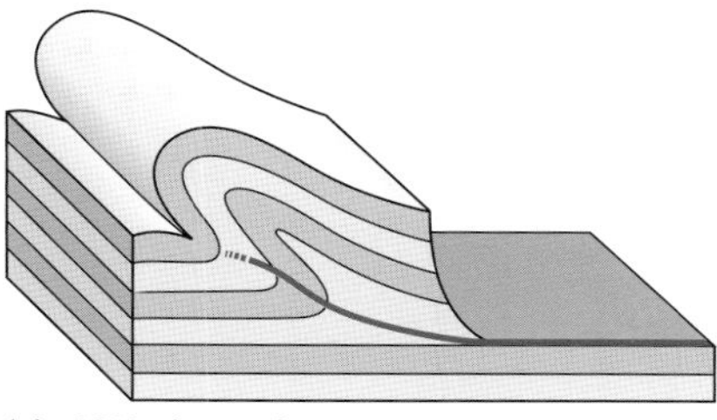

(c) 48% shortening

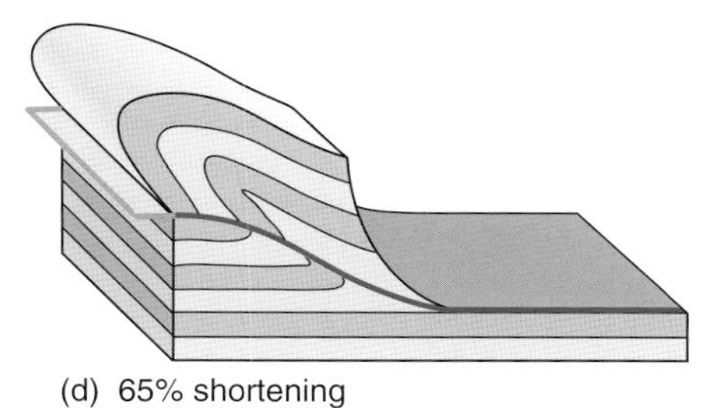

(d) 65% shortening

Figure 11.6 Cartoons showing stages in the development of a fold pair related to a thrust tip. (a) At first, ductile thickening ahead of the thrust tip is important, causing folds to form in the hangingwall. (b) Asymmetric folds form, with a sense of overturning consistent with the direction of hangingwall movement. (c) The thrust propagates through the overturned limb of the fold. (d) Finally, the anticline in the hangingwall is carried forward as the thrust tip advances beyond the fold.

11.4.3 Buckle folds

Folds can also form as a mechanical consequence of shortening across layers, completely independent of fault movements. If a competent layer set in a less competent matrix is shortened along its length, a mechanically unstable situation develops where the competent layer is deflected sideways and develops into a fold. These folds are called **buckle folds**, because they have formed by buckling – shortening along the length of the layers. They need not be directly related to faulting, and are usually they are not. You can make a buckle fold very simply by holding the two ends of a plastic ruler and moving your hands towards each other. (Contrast this with a bend fold, which you make if you bend the ruler over the edge of a desk.)

Figure 11.7 illustrates the principles of buckle folding. Suppose a layered composite of materials of different competences shortens along its length. If competent layers are separated from other competent layers by a considerable thickness of incompetent material, the wavelength of the buckle fold is controlled by both the thickness of the buckled layer and the competence contrast between layers. An increase in the thickness of the competent layer or an increase in the competence contrast both lead to an increase in the wavelength of the fold.

Where the competence contrast is high, as it is between layers (w) and (z), buckle folds develop with long limbs and long wavelengths. Where the competence contrast is low, between layers (y) and (z), the folds have very short limbs. The idealized situation shown in Figure 11.7a can be duplicated in laboratory experiments (Figure 11.7b), and also seen in real rocks (Figure 11.7c).

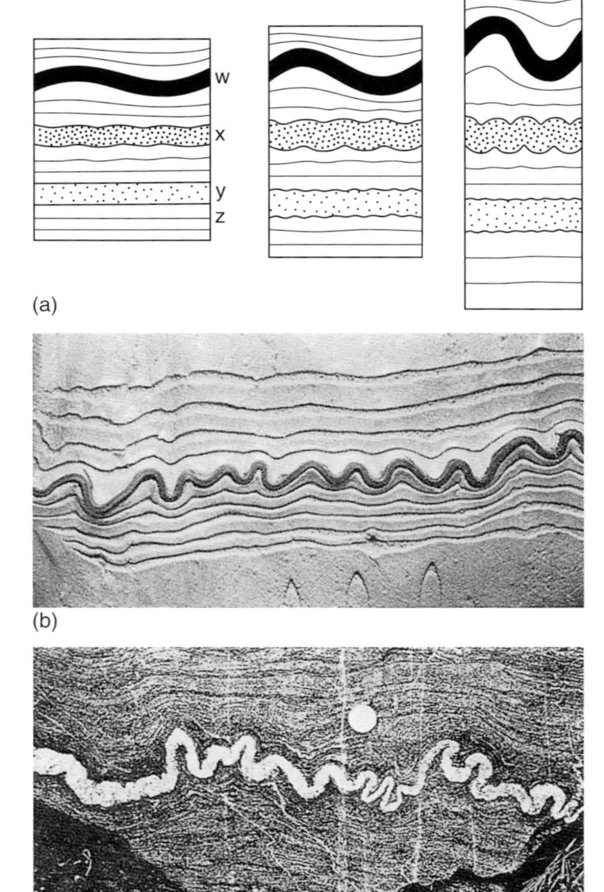

Figure 11.7 Development of buckle folds. (a) Diagrammatic cross-sections through buckle folds in materials with a competence contrast (competence of materials: w>x>y>z; z is the matrix); (b) a laboratory experiment with layers of different competence in plastic modelling materials; and (c) a real example from metamorphic rocks in Zimbabwe.

- ❏ How does the orientation of the axial planes of these buckle folds relate to the direction of shortening?
- ■ The fold axial planes lie approximately at right angles to the original layering, and the shortening direction is parallel to the initial layering. So the folds are oriented with their axial planes approximately at right angles to the direction of shortening.

This question illustrates an important point about buckle folds; they are a ductile way of shortening a layered succession of rocks. In examples where it is clear that folds have formed by buckling, and where independent evidence exists to determine the overall shortening direction, we find that folds are commonly oriented such that their axial planes lie roughly at right angles to the shortening direction.

11.4.4 COMMON FOLD PROFILES

Whether they are generated through fault-related processes or by buckling, we can consider folded layers in terms of two commonly seen fold profiles. Figure 11.8a shows a cross-section through a **parallel fold**. In parallel folds, the layer keeps a constant thickness right around the fold. This is possible only because the radius of the outer arc of the fold is significantly greater than the radius of the inner arc. Contrast this profile with that in Figure 11.8b, which shows a cross-section through a **similar fold**. In similar folds, the shape of two neighbouring folded surfaces is almost identical. To achieve this, the thickness of the folded layer has to vary; it is much thicker in the fold hinge than it is in the fold limb.

Parallel and similar folds are not extremes. You may see folds where the folded layer thins into the crest, or folds where thickening into the crest is so marked that the shape of adjacent folded layers is no longer identical. We highlight these particular fold shapes because natural folds are often a combination of parallel folds in competent layers and similar folds in incompetent layers. Parallel folds maintain bed thickness, but they cannot be stacked together indefinitely because of their rapidly changing arc radius. Similar folds can be stacked indefinitely, but require dramatic changes in bed thickness (and therefore rock volume) to develop. Natural folds are often a compromise between the two. In reality, the competence contrast between a buckling layer and its matrix is the most significant factor – high competence contrasts give rise to parallel folds whilst low competence contrasts produce similar folds.

(a)

(b)

Figure 11.8 Cross-sections through (a) parallel and (b) similar folds.

11.5 How tectonic fabrics form

We have just seen that folds owe their existence to heterogeneities within the rock succession. Whether they form by buckling or by fault-related processes, competence contrasts between different lithologies must be present before folds can develop. What happens to uniform, homogeneous rocks as they deform?

11.5.1 Homogeneous lithologies and slaty cleavage

When a homogeneous rock, such as a mudstone, is deformed, it will shorten and thicken rather than generate folds or faults. How exactly does this happen? Mudstones are composed of many tiny grains of platy minerals, and depending on the exact rock composition, generally a percentage of minute quartz (or calcite) grains as well. Each of those individual grains must respond to deformation. Mineral grains deform in three ways; by distortion, by rotation coupled with translation, and by dissolution and regrowth. When grains are small, as they are in uniform mudstones, the last two are the more important processes.

Let us consider first what happens to platy grains during deformation. Platy grains, such as micas and clay minerals, are mostly sheet silicates which are composed of layers of silicate groups (Block 2 Section 4.5). Assume that the tiny mineral 'sandwiches' were originally randomly orientated in the undeformed rock (Figure 11.9a). As the rock shortens, almost all of these grains will rotate. Exceptions will be grains already oriented with their longest dimension either parallel to or normal to the shortening orientation. As shortening (and attendant lengthening) of the rock progresses, platy grains will rotate first into a crude parallelism (Figure 11.9b) and onwards into an almost perfect alignment (Figure 11.9c).

At the same time as physical rotation is taking place, increasing temperatures and pressures during deformation are driving mineralogical changes. Sheet

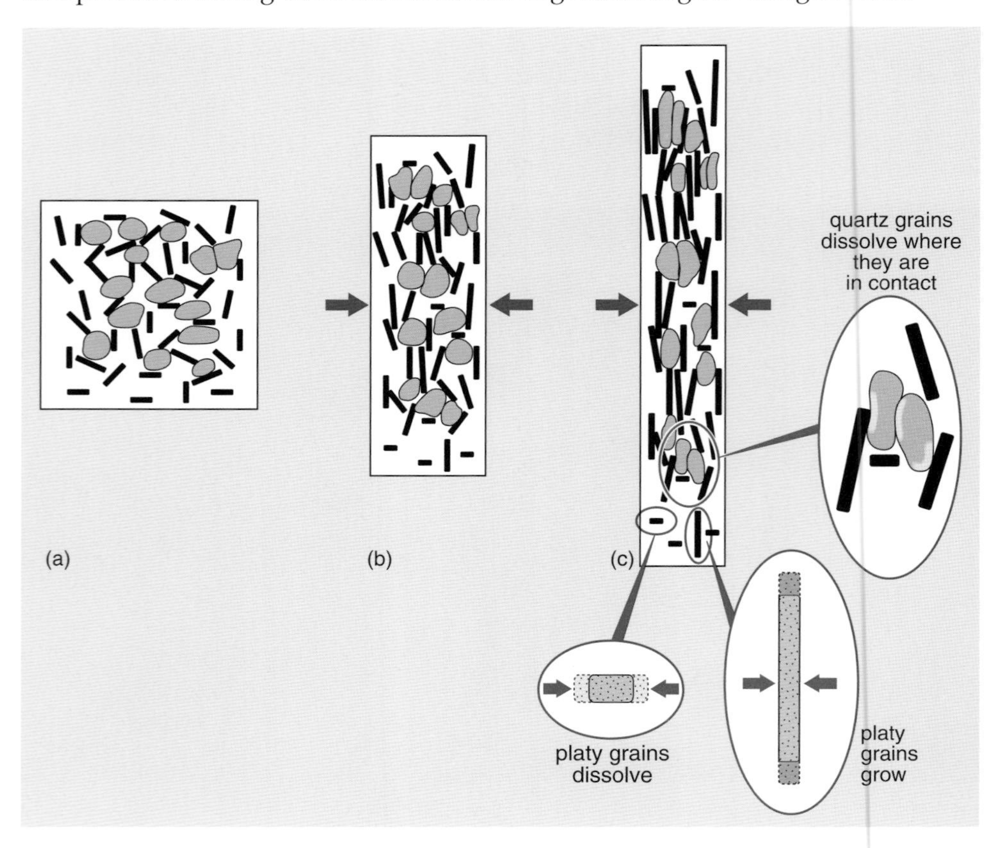

Figure 11.9 Stages (a)–(c) in the development of a cleavage by shortening in mudrocks through grain alignment and recrystallization. The rod-like shapes represent platy minerals, the more equant grains represent quartz. Red arrows indicate shortening direction.

silicates dissolve and recrystallize by adding or subtracting 'building blocks' from the ends of their plate lattice, rather than by adding or subtracting more plates. Whether any individual lattice grows or dissolves depends on its orientation relative to the shortening direction. Grains that lie parallel to the shortening direction have their ends highly stressed, so they dissolve. The phenomenon is known as **pressure dissolution** (usually referred to as pressure solution). Conversely, grains at right angles to this are in an ideal orientation to grow. Grains lying with their long axes at right angles to the shortening direction will grow at the expense of those lying parallel to shortening (Figure 11.9c).

However, the effects of pressure dissolution on small quartz or calcite grains within the rock are usually much more significant. These minerals are relatively equidimensional, and tend to dissolve away at points where they are in contact with other minerals (Figure 11.9c). This process removes material from the quartz or calcite crystal into solution, reducing the size of the grain, and allowing nearby platy minerals to rotate still further. Quartz or calcite is reprecipitated from solution, either onto the low-stress areas of existing grains, or into intergranular sites as new minerals, or as quartz or calcite veins within the rock.

The combined effect of these processes of rotation and dissolution enlongates grains physically and chemically. The end-product is a rock with a very strong alignment of grains, and which splits preferentially along the plane defined by the oriented plates. It has *slaty cleavage*. A cleavage that forms dominantly by the process of pressure dissolution is commonly called a **pressure solution cleavage**. Many analyses of strain in slates have shown that slaty cleavage forms perpendicular to the direction of overall maximum shortening, which can be in excess of 50%. Volume reduction through pressure dissolution is also an important process during slaty cleavage formation.

It is rare that vast thicknesses of sedimentary rocks are completely uniform in composition, however. Even in predominantly muddy successions, there are usually some beds which are sandier. Lithological variation provides competence contrast. This means that in successions of, say, mudstones with interbedded sandy beds, cleavage formation will be accompanied by buckle folding. Fold style will be controlled by the thickness and competence contrast of the competent lithologies. Cleavage can and usually does form in all lithologies, but slaty cleavage is always better developed in the muddy, incompetent layers than it is in the sandier, competent layers.

The cleavage often also changes its orientation from layer to layer. In folded competent beds, rather than being exactly parallel to the fold axial planes, the cleavage typically converges *downwards* towards the axial plane of an anticline. In less competent beds, the cleavage 'fan' converges *upwards* (Figure 11.10). Differential orientation of the cleavage plane from bed to bed is known as **cleavage refraction**.

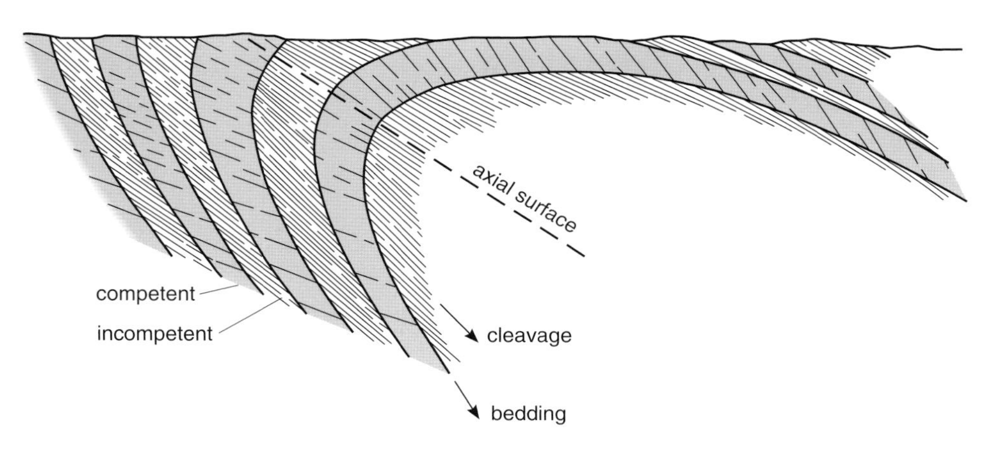

Figure 11.10 Cleavage refraction in folded layers of different competence. Coloured beds are competent sandstones; unornamented beds are incompetent mudstones. The slaty cleavage orientation is shown by broken lines that are nearly parallel to the axial plane (or axial surface) of the fold.

It is important to realize that folds and cleavage are *independent* ways of achieving shortening in rocks. The axial planes of folds and cleavage planes both lie at right angles to the maximum shortening direction. Many folds carry an axial-planar cleavage, but not all folds do. There are also places where the rock is cleaved, but not folded (for example, the slate quarries in Valencia Island that you saw in Video Band 8).

11.5.2 METAMORPHIC FABRICS

So far, we have considered tectonic fabric development in sedimentary rocks, but metamorphic rocks also almost always carry tectonic fabrics. Metamorphic rocks often have very complex histories, extending over long periods of time. Not all the minerals present in the rock necessarily formed at the same time. Yet by looking closely at the texture of a metamorphic rock, to see the relative ages and orientations of its metamorphic minerals, we can usually unravel the history of changing pressure and temperature conditions, and relate these to the rock's deformation history.

In Block 2 Section 8.2, we noted that deformed platy minerals like micas tend to form a foliation whereas elongate minerals like amphibole tend to form a lineation. These are textures; the collective term for the range of metamorphic textures that can form is the rock *fabric*.

The presence of minerals showing a foliation or lineation in a metamorphic rock shows that the rock must have been deformed either during or after mineral growth. Several different mechanisms can account for the alignment of minerals in a metamorphic rock, yet each is a means by which the minerals, and therefore the rock itself, can be shortened.

First, deformation can occur *within* individual grains through the movement of *dislocations* in the atomic lattice (Block 2 Section 2.4). The result is the elongation of the grain shape at right angles to the shortening direction. Lattice distortions because of deformation are common in quartz, and can be recognized by wavy *extinction* when the rock is viewed in thin section between crossed polars (Activity 6.2 of Block 2).

Secondly, deformation during metamorphism usually causes crystallization and growth of new minerals in a preferred orientation. This can occur through nucleation of minerals with a common alignment, because crystals require less energy to nucleate and grow in certain orientations with respect to the shortening direction. For example, platy minerals tend to grow with the flat sides of their crystal faces at right angles to shortening. The rock develops a cleavage and elongate minerals grow with their long axes in the cleavage plane.

Recrystallization of existing minerals without a platy shape, such as quartz, results in elongated grains through a two-stage process. First, grains break down into smaller grains, known as **subgrains**. Secondly, these subgrains are rearranged by translation and rotation into an elongated ribbon (Figure 11.11). A *ribbon texture* in quartz grains is characteristic of high-strain rocks known as *mylonites* (Plate 8.3b), although we must emphasize that not all metamorphic rocks that show the development of a ribbon texture are mylonites.

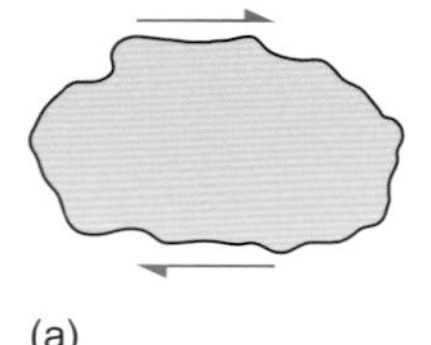

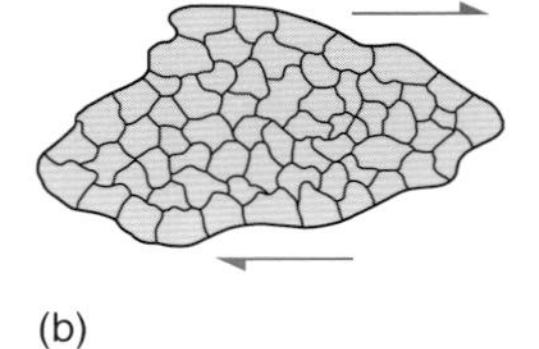

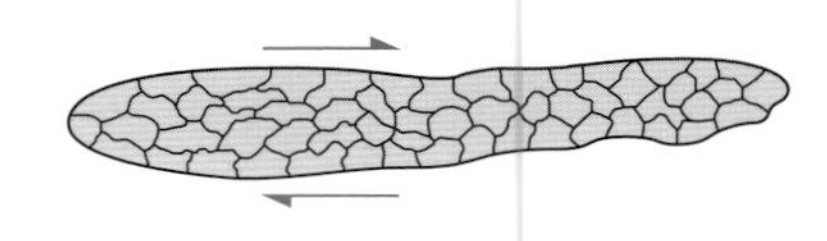

Figure 11.11 Stages in deformation of quartz grains. (a) Original grain subjected to deformation; (b) breakdown into subgrains; (c) reorganization of subgrains to form ribbon texture.

However, deformation does not always occur exactly at the same time as metamorphism. In fact, it is usually possible to determine the relationship between the time of deformation and the time of metamorphism by examining the fabric of the rock. For example, if metamorphism occurs before deformation, then early-formed metamorphic minerals will themselves be deformed. This is clearly shown by the distorted mica flakes seen in Plate 11.1.

Metamorphism must have occurred early in a sequence of deformation events if new minerals are aligned to form a planar schistosity, and a later episode of deformation affects this schistosity (Plate 11.2). The result is a shortening of the layers and fold formation. In Plate 11.2, the rock has been shortened along the horizontal axis of the photograph.

Where metamorphism occurs after deformation, new metamorphic minerals will not be aligned, but instead grow across the fabric of the rock. This often happens during contact metamorphism, when hot magma intrudes rocks that may already have a fabric, causing new minerals to grow. In Figure 11.12, you can see that a cleavage in a slate has been overgrown by well-formed rectangular crystals of andalusite – a mineral characteristic of contact metamorphism. In this case, deformation must have preceded contact metamorphism. This type of texture is also common in regionally metamorphosed rocks, and indicates continuing elevated temperature and pressure, and thus mineral growth, after deformation.

Most commonly however, metamorphism and deformation occur at much the same time. The relationship between them can sometimes be evaluated by examining trails of inclusions within metamorphic porphyroblasts. For example, a garnet growing in a rock which already has a planar schistosity (Figure 11.13a, overleaf) will grow around inclusions that lie parallel to the schistosity. Subsequent deformation (Figure 11.13b, c) may cause the porphyroblast to rotate, and if it continues to grow the garnet will engulf inclusions around its rim at an angle to the inclusions in its core (Figure 11.13d). The final result is an S-shaped or sigmoidal pattern of inclusion trails characteristic of mineral growth during deformation (Figure 11.13e).

Figure 11.12 Porphyroblasts of andalusite in a slate from a thermal aureole (plane-polarized light; width of image = 3.5 mm).

(a)

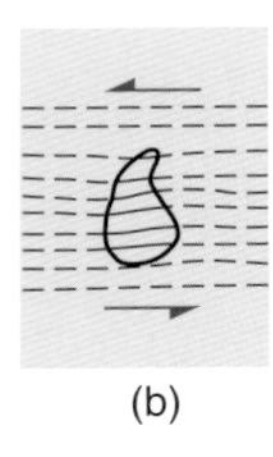
(b)

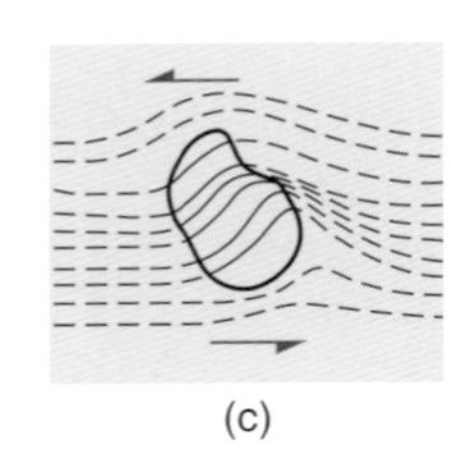
(c)

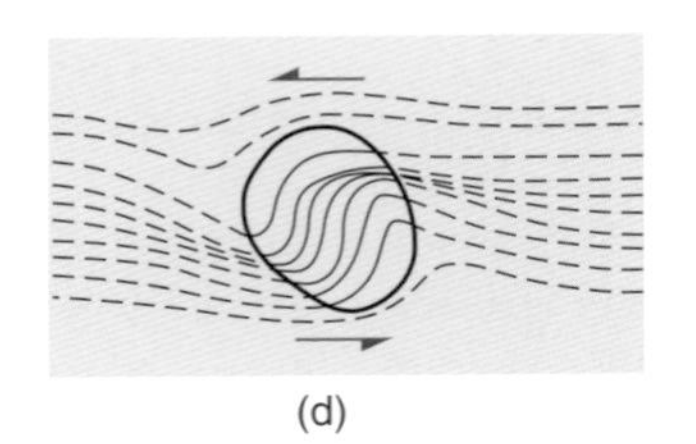
(d)

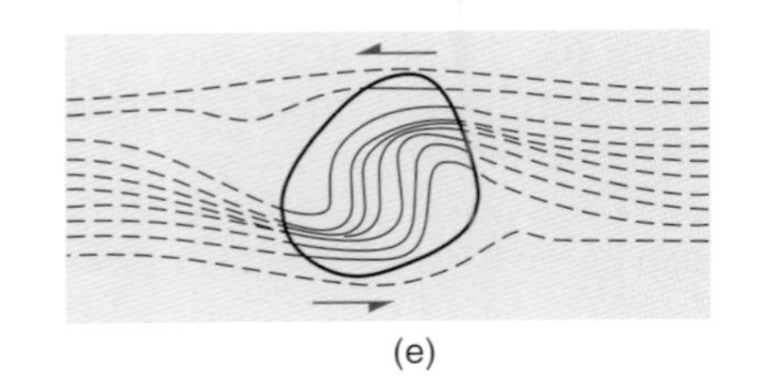
(e)

Figure 11.13 Sequence by which S-shaped inclusion trails develop in garnet that is rotated during deformation.

Activity 11.4

The analysis of fabrics of metamorphic rocks is really the application of simple rules, based mainly on common sense, to the fabrics we see down the microscope. Try Activity 11.4 now, to explore two contrasting metamorphic fabrics resulting from regional metamorphism.

11.6 TIMING ROCK DEFORMATION

To end this Section, we touch briefly on how structural geologists date rock deformation. This information is necessary if we are to know how long any deformation episode lasted, or when an episode of deformation or metamorphism took place. The clearest indication of the age of deformation is given by the age of the rocks themselves. Rocks cannot deform before they came into existence, so if the original age of a deformed rock is known, then the *maximum* age of deformation can be established. Similarly, undeformed sedimentary rocks that lie unconformably on deformed rocks must be younger than the deformation, as must undeformed igneous rocks that are intruded into deformed rocks.

❑ From Block 1 Section 4.6.1, what is (i) the youngest and (ii) the oldest age for deformation of Carboniferous strata in the Mendips area?

■ (i) Since we established in Block 1 Section 4.6.1 that these Carboniferous rocks are unconformably overlain by unfolded Triassic rocks, deformation must pre-date Triassic times.

(ii) We can see from the Ten Mile (South) map that Upper Carboniferous rocks are folded, so folding must have happened after they were deposited. The oldest age for folding is therefore late Upper Carboniferous times.

In recent years, it has become possible to perform radiometric dating on individual minerals such as mica or garnet. If it is clear from the rock fabric that the mineral grew before, during or after deformation, then dating the mineral puts precise limits on the age of deformation.

There are other ways in which the relative age of deformation can be established, albeit less precisely. It may be that both the limbs and axes of folds appear themselves to have been folded; these structures are often called **refolded folds**. The upright, tight-to-isoclinal fold in the lower right corner of Figure 11.14a has been folded by a later upright open fold – if you sketch in the axial plane of the tight fold, you can confirm that it has been folded. When this happens, we can be sure that the folds *with folded axial planes* belong to a relatively early episode of deformation, which pre-dates at least one later folding episode. Although we know nothing of the exact age of either episode of folding (except that they are both younger than the rocks themselves), we do know the relative order in which the two fold episodes came.

This type of observation can be extended to include folded cleavages. Folds of cleavage must indicate two distinct episodes of deformation – the first to form the cleavage and the second to generate folds in it (Figure 11.14b).

(a)

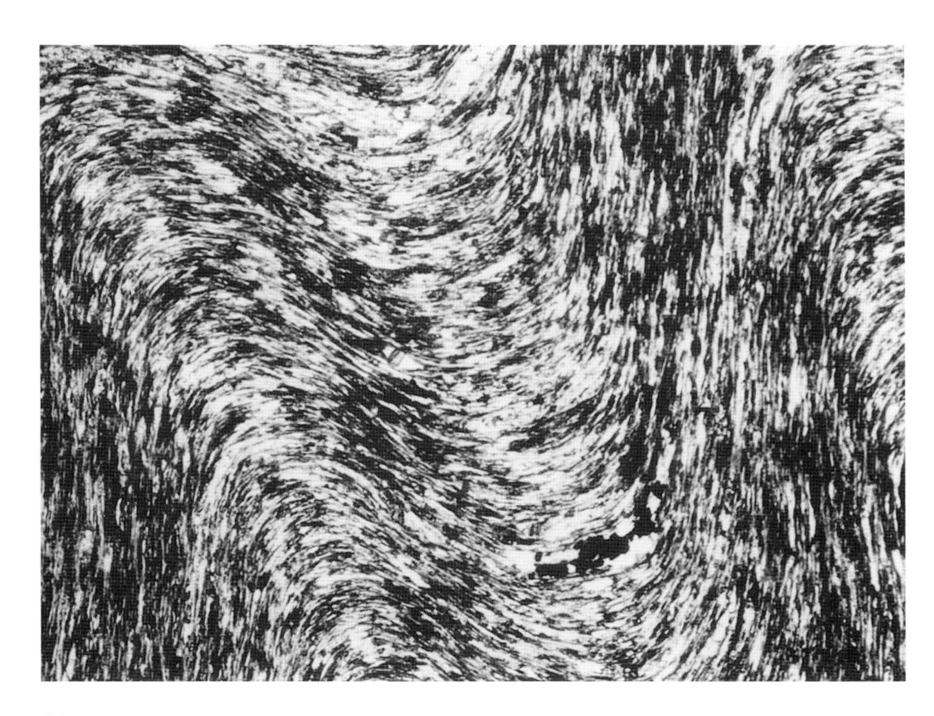

(b)

Figure 11.14 (a) Refolded folds. The upright tight-to-isoclinal fold in the lower right corner has been folded by a later upright open fold. The area shown is about a metre across. (b) A photograph of a thin section (a few mm across) showing folded cleavage. Note the strong alignment of both light and dark grains.

11.7 Summary of Section 11

- How rock responds to stress depends on the temperature, pressure and rate at which it is deformed, the properties of the rock material itself, and many other variables.
- Rock deformed near the surface of the Earth shows more evidence of brittle failure than that deformed at depth.
- Rock deformed at high strain rates shows more brittle behaviour than that deformed at low strain rates.
- Rock composition has a profound effect on deformation. Competent lithologies are stiffer, flow less easily and break more readily than incompetent lithologies.
- Sandstones, limestones and most igneous rocks usually deform in a competent manner, whilst mudstones, shales and evaporites generally deform in an incompetent manner.
- Joints are attributed to brittle failure, often caused by uplift and erosion.
- The orientation of rock layers often controls the orientation of the dynamic stress field, hence the orientation of faults.
- Normal faults form in response to horizontal stretching, often at high angles to sub-horizontal bedding. Reverse faults form in response to horizontal shortening, often at low angles to sub-horizontal bedding. Strike–slip faults allow the crust to change its lateral dimensions.
- The characteristic profile of faults, particularly thrusts, is one of flats and ramps. Flats lie parallel or almost parallel to beds, whilst ramps cut up through beds
- Fault-related folds form as a mechanical consequence of fault displacement. Buckle folds form as a mechanical consequence of shortening along layers.
- In parallel folds, the layer thickness remains constant around the fold. In similar folds, the shape of two neighbouring surfaces is identical, but the bed is thicker in the fold hinge than in the fold limb.
- When homogeneous lithologies deform, grains rotate physically, dissolve and regrow. The end-product is a rock with a fabric characterized by very strong alignment of platy grains, i.e. with slaty cleavage.
- Cleavage and folding are independent ways of shortening a body of rock.
- Several mechanisms account for the alignment of platy or elongate minerals in a deformed rock. Deformation can elongate grains through the movement of dislocations in the atomic lattice. Nucleation of new minerals with a common orientation may result in the rock developing a cleavage. Recrystallization of minerals that do not have a platy shape, particularly quartz, can result in the breakdown of grains into subgrains.
- Where metamorphism occurs after deformation, new minerals often overgrow the fabric of the rock.
- A clear indication of the age of deformation is given by the age of the rocks themselves. If the original age of deformed rock is known, the earliest time of deformation can be established. Undeformed rock must post-date deformation. Refolded folds and folded cleavage indicate two or more distinct episodes of deformation.

11.8 Objectives for Section 11

Now you have completed this Section, you should be able to:

11.1 List the physical parameters that control whether rock deforms in a brittle or a ductile way.

11.2 Identify competent and incompetent lithologies from their behaviour within structures.

11.3 Explain how joints form, and why normal faults are usually steep whilst thrusts are usually low-angle structures.

11.4 Describe the separate processes by which folds form.

11.5 Explain why homogeneous rocks carry a slaty cleavage.

11.6 Describe the mechanisms by which simple metamorphic fabrics are formed.

11.7 Interpret simple metamorphic textures as seen in exposures, hand specimens and thin sections in terms of relative timing of deformation and metamorphism.

11.8 Determine the age of rock structures from geological evidence.

Now try the following questions to test your understanding of Section 11.

Question 11.2 Which of these structures – normal faults, ductile shear zones, thrust faults, joints – would you expect to form at depth and which in the upper few kilometres of the crust? (*Explain your reasoning in 30–40 words*)

Question 11.3 Suggest a way in which (a) a **homogeneous** and (b) a **heterogeneous** body of rock could shorten in a ductile fashion. Which structures would characterize each of these processes? (*About 100 words*)

Question 11.4 Identify the main differences between parallel and similar folds. (*About 100 words*)

12 TECTONIC ENVIRONMENTS

'In order to make this land a permanent body, resisting the operations of the waters, two things had been required; first, The consolidation of masses formed by the collections of incoherent materials; secondly, The elevation of those consolidated masses from the bottom of the sea ... to the stations in which they now remain above the level of the ocean.... By what power of nature the consolidated strata at the bottom of the sea had been transformed into land.'

From the 1785 Abstract of James Hutton's *Theory of the Earth*.

James Hutton's words introduced the first Earth Science revolution – an understanding that the Earth had not simply been created exactly as it now is, but that it had evolved into this form by processes that could be observed and understood. Two hundred years, later Earth Science had to come to terms with its second revolution – plate tectonics. The success of plate tectonics lies in the ability to link together previously unrelated geological, geophysical and geochemical observations. Hutton would no doubt have appreciated seeing the unification of many geological processes within the framework of moving and interacting plates.

Ironically, James Hutton's legacy – the fact that 'the present is the key to the past' – appears to be less useful to structural geologists than it is to many other Earth scientists. Rock deforms at depth under considerable confining pressures and temperatures; conditions which can only be duplicated in the laboratory for relatively small samples, not for whole mountain belts. Rock also deforms very slowly, which means that within human life-spans we simply cannot wait long enough to observe most rock-deforming processes taking place at anything like their natural rates.

However, if the present *is* the key to the past, we must try to interpret structures we see in old rocks in terms of plate tectonics. The processes that happen at plate boundaries where plates interact – continental breakup, separation and drift, and continental collision – must dictate the structural style of most deformed sedimentary and metamorphic rocks.

Not every tectonic setting shows every type of structure. For example, multiple generations of tight folds superimposed one upon another are much more characteristic of high-grade metamorphic rocks than of low-grade sedimentary deposits. Yet irrespective of rock age, time of deformation, and global location, certain structures and structural histories are repeatedly found together. Worldwide, and throughout geological time, there appear to be four principal tectonic *associations*.

First, there are areas of continental lithosphere that have been tectonically inactive or almost inactive for hundreds of millions of years. Such areas are called **cratons**, after the Greek word *kratos* meaning 'strength'.

Secondly, there are areas of continental lithosphere characterized by thick successions of sedimentary rocks, relatively simple structures dominated by normal faults, subsidence over time, and a conspicuous lack of igneous and metamorphic rocks. We term these areas **sedimentary basins**.

Thirdly, there are belts containing tectonically interleaved sedimentary and metamorphic rocks, and igneous (particularly granitic) intrusions. These belts are characterized by uplift, and by complex structures – major thrust faults and large-scale folds with cleavages. Typically, such belts form mountain chains, or at least the eroded remnants of former mountain chains. They are known as **orogenic belts**, from the Greek *oros* meaning 'mountain' and *genesis* meaning 'production'.

Fourthly, there are smaller yet significant areas of continental lithosphere which are not necessarily mountainous and may contain sedimentary, igneous or metamorphic rock, but which are characterized by large-scale lateral rather than vertical movements. These are known as **strike–slip zones**.

❑ How do the oceans, floored mainly by mafic igneous rocks, fit into this scheme?

■ Badly; these four associations classify tectonic patterns in *continental* rather than *oceanic* crust.

12.1 STRUCTURES ASSOCIATED WITH CRATONS

The biggest cratons are several thousand kilometres across, though they are highly variable in size and shape. They are usually situated well away from present-day plate boundaries. They are stable areas – cratons show no evidence of significant tectonic activity over long periods of geological time.

Typically, they contain a succession of sub-horizontal sedimentary strata, resting unconformably on ancient metamorphic rocks. Over vast areas, these sedimentary rocks are not significantly folded or faulted, and frequently not even tilted; often the only tectonic structures are joints. The succession of strata may record deposition over many hundreds of millions of years. Often the sequence of rocks is relatively thin – hundreds rather than thousands of metres thick.

Cratons generally show evidence for variations in the rate of subsidence and/or uplift over time, primarily due to changes in *isostatic equilibrium* (Section 2.2.3). Figure 12.1 shows a view of the North American craton at the Grand Canyon in Arizona, USA. Here the Colorado River cuts down through about 1600 m of sub-horizontal sedimentary rocks which lie unconformably on metamorphic 'basement' rocks over 700 Ma old. The strata exposed at the present surface are

Figure 12.1 Grand Canyon, USA. Vast areas of almost horizontal sedimentary rocks like these are characteristic of tectonically stable areas of the continents.

sediments about 250 Ma old. The plateau is a remnant of a large uplifted region in the western USA where it is believed about 3 km of younger strata that were once deposited have been eroded during the past 20 Ma.

❑ How could increased heat flow under the craton have led to this uplift?

■ Increased heat flow from the mantle would cause thermal expansion that would buoy up the lithosphere isostatically and produce uplift.

In plate tectonic terms, cratons show that deformation is rare well away from active plate boundaries. Cratons predictably show few tectonic structures. Major changes happen only if the plate is disturbed by some event sourced from outside the plate itself, like for example the effects of a deep-seated mantle plume (Section 7.2).

12.2 STRUCTURES ASSOCIATED WITH SEDIMENTARY BASINS

Sedimentary basins are areas of continental crust characterized by thick successions of sedimentary rocks deformed by arrays of predominantly normal faults. We know (i) that normal faults are caused by stretching parallel to the Earth's surface, and (ii) that thick accumulations of sediment can only be accommodated where the surface is actively subsiding. These two features suggest that basins form by regional surface-parallel extension accompanied by thinning normal to the Earth's surface.

❑ What structures would you expect to see near the surface in sedimentary basins?

- ■ Near the surface, brittle structures would dominate. Normal faults would be common, and horsts and grabens may form. Sedimentary successions may be extensively jointed.

Figure 12.2 shows a structural cross-section through part of the North Sea basin, the largest sedimentary basin in the UK. In the centre of the basin, up to 8000 m of sedimentary rocks have been deposited in the 250 Ma since Triassic times, cut by large normal faults throwing down towards the Viking Graben, which lies in the centre of the basin. These large normal faults separate tilted blocks some 10 km wide.

- ❑ What would be the implications, in terms of brittle and ductile structures, of surface-parallel extension deeper down, below the section in Figure 12.2?
- ■ At depth, ductile behaviour would be expected, probably expressed both as ductile shear zones and ductile thinning.

Just how the North Sea basin formed is shown diagrammatically in Figure 12.3. When the lithosphere thins because of crustal stretching, the upper surface must subside. The base of the crust (marked by the Moho) rises, bringing the hot mantle nearer to the Earth's surface (Figure 12.3b). The result is an increase in heat flow through the stretched zone and an associated buoyant effect caused by isostatic disequilibrium. If the stretching rate is fast, the buoyant effect of high heat flow will be significant, and may cancel out any surface subsidence due to crustal thinning.

When extension stops, faults become inactive. The upper layers of hot lithospheric mantle cool, and heat flow wanes (Figure 12.3c). As they cool, they become denser and sink isostatically, lowering the surface of the stretched area. This space will become filled with sediments in which faulting has not occurred. The basin will have two structural components: a lower one containing faulted and rotated blocks of older sediment which represents the *extension* phase; and an upper one containing unfaulted, younger sediment which represents a *thermal collapse* phase. This process has happened in the North Sea, as we can clearly see by comparing Figure 12.2 with Figure 12.3c.

- ❑ Looking at Figure 12.2, by when had most of the faulting and block rotation taken place?
- ■ By Tertiary times, because Tertiary rocks are not affected by faulting whereas the older ones are.

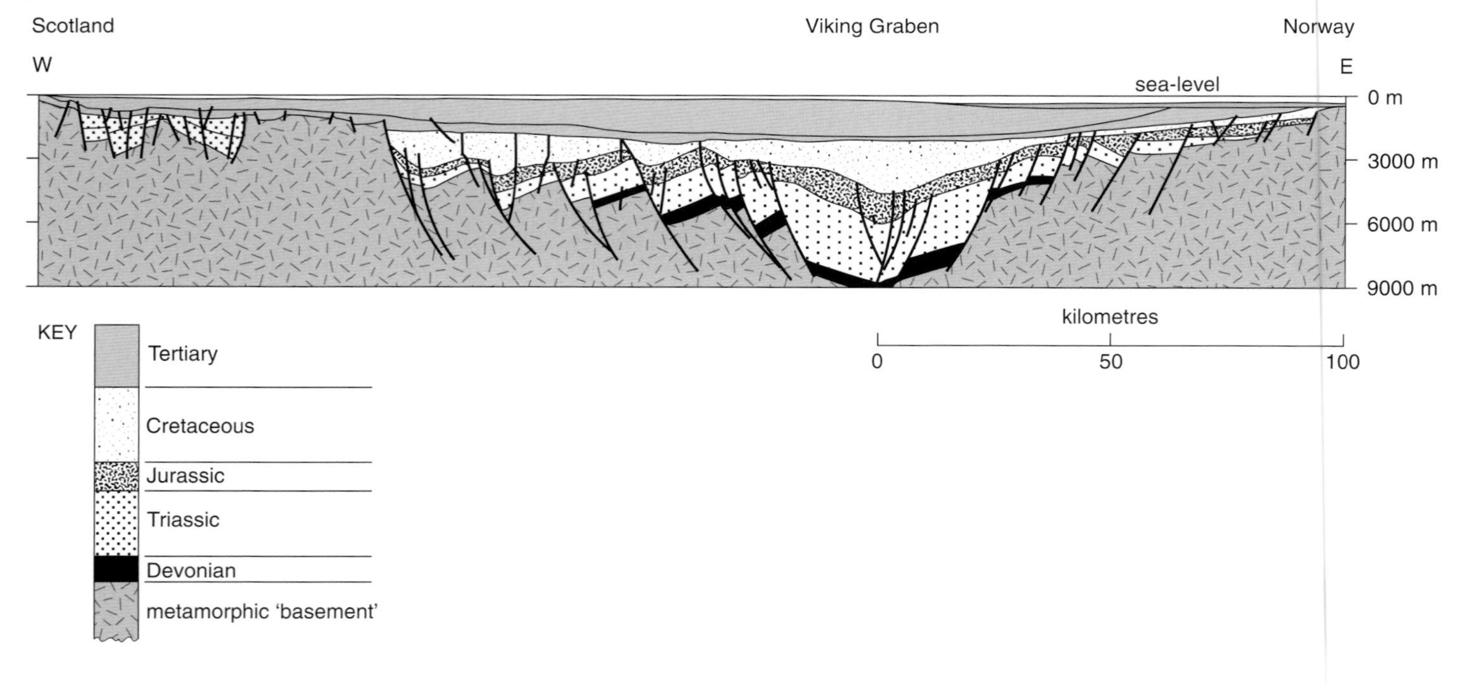

Figure 12.2 A cross-section through the upper crustal layers of the northern North Sea sedimentary basin. Note the scale of this basin, extending for more than 200 km between Scotland and Norway. Heavy black lines are faults. Vertical scale is × 5 greater than the horizontal scale.

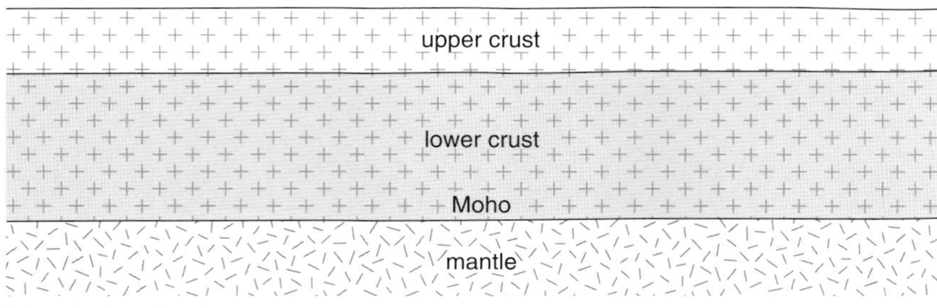

(a) prior to extension

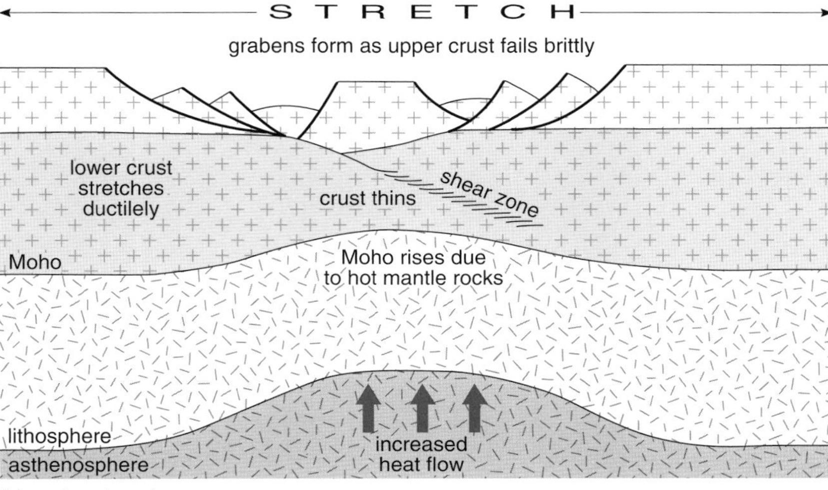

(b) extension phase

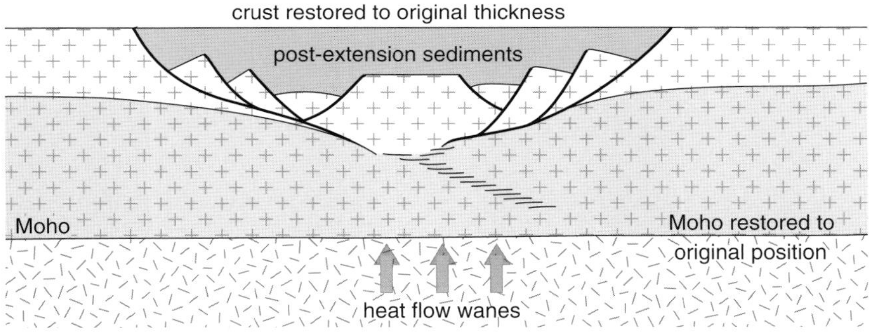

(c) thermal collapse phase

Figure 12.3 Formation of a sedimentary basin by extension followed by thermal collapse. (a) The lithosphere prior to stretching. (b) The lithosphere thins as it stretches, and its upper part fails in a brittle manner, giving rise to rotated blocks separated by normal faults. Heat flow increases because hot mantle rocks are near the surface, and thermal expansion leads to isostatic uplift. (c) When stretching stops, heat flow wanes and the basin subsides, again due to isostatic disequilibrium. More sediment is deposited. Note the two structural components of the basin: an upper unfaulted part and a lower faulted part.

12.2.1 THE PLATE TECTONIC SETTING OF SEDIMENTARY BASINS

If crustal stretching is extensive enough, and if extensional structures penetrate deeply into the crust, magma may rise along the fault and shear zones. Volcanoes and associated intrusions form, such as those you saw in Video Band 5. If crustal stretching continues still further, basaltic magma will be injected into continental crustal rocks, usually along sub-vertical sheeted dyke complexes, and if erupted under water may form pillow lavas. When stretching has developed this far, new oceanic crust is forming within stretched continental crust. A zone that started life as a sedimentary basin has become a site of continental separation.

In past tectonic cycles, there have been frequent examples of two continents separating along a zone of crustal extension, as shown in Figures 2.9 and 12.4. Extension is initiated (Figure 12.4a), and as separation continues, the rising

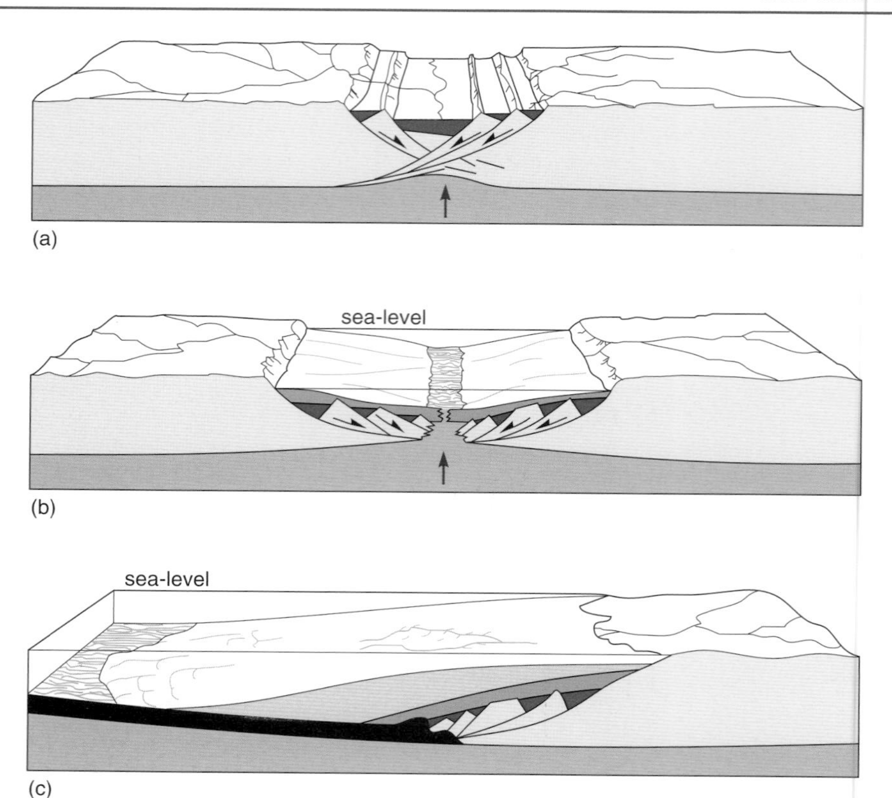

Figure 12.4 Cartoon cross-sections of the sequence of events at a divergent continental plate boundary. At first (a), the upper parts of the crust extend by developing a series of brittle normal faults. The continental surface also sinks, creating a basin which provides a site for the accumulation of sedimentary or volcanic rocks (darker red). As the plates continue to diverge (b), the lower lithosphere will rise and melt. Oceanic crust forms in the centre of the basin and the continent has separated into two. Eventually (c), the thinned margins subside as they cool, allowing a further cover of unfaulted sediments to be deposited (lighter red). A passive continental margin has been created.

lithospheric mantle starts to melt. Volcanic rocks become increasingly important within the basin (Figure 12.4b). The process ends in the separation of two bodies of continental crust and the formation of a new zone of oceanic crust between. As each continental margin moves away from the new spreading axis, it cools and subsides. The continental margin becomes blanketed by sedimentary rocks (Figure 12.4c). The edge of the continental crust is now a *passive continental margin*, and spreading has become focused at a mid-ocean ridge.

Most large sedimentary basins that developed in the past were either (i) intracontinental graben systems, or (ii) passive margins that lay between oceanic and continental crust as a result of continental separation. These latter sites are the very places that are destined eventually to become involved in continental collision.

12.3 Plate tectonics and orogenic belts

If sedimentary basins and passive margins form at the start of the continental plate tectonic cycle, how does the cycle end? Continents cannot continue to separate indefinitely.

- ❑ Why not?
- ■ Because the globe is finite in size. New oceanic lithosphere created at a constructive plate boundary must be compensated for by subduction at a destructive plate boundary.

Separating continental fragments, each trailing a passive margin, are at the same time moving towards other continental crustal fragments. Eventually the continents must meet.

We have a rough idea of the time-scale over which this happens. Except for ophiolites, those rare parts of oceans that have become 'fossilized' within continental crust (Section 7.1), the oldest oceanic crust on the globe today is about 200 Ma old. All older oceanic crust has been subducted at convergent plate boundaries. The implication is that after about 200 Ma passive continental margins must convert to active, destructive plate boundaries. Since it took about 200 Ma to start and form an ocean basin fully, then oceans should close and continents should converge within roughly 400 Ma. We believe that plate tectonics has been operating on Earth for some 4000 Ma. There has been enough time for many cycles of separation of a mega-continent into fragments, leading to ocean formation and continental drift, leading to continental collision and mega-continent reconfiguration.

We know, for example, that the Indian continental fragment separated from a giant Southern Continent about 115 Ma ago, and collided into Asia between 55 and 40 Ma ago. In about 70 Ma, it journeyed a quarter of the way around the globe. A host of Himalayan igneous and metamorphic rock ages of between 40 Ma and the present time suggest that there is strong correlation between the impact of India into Asia and the formation of the Himalayan mountain belt.

Colliding continents form orogenic belts. Which structures characterize colliding continents? We can start to answer this question by making three predictions. First, passive continental margins have a strongly extensional form, so continental margins approaching subduction zones are also likely to have a strongly extensional form. We can predict that any structures that form during collision will be superimposed onto pre-existing extensional structures (Figure 12.5a).

Secondly, as two continents meet and impact together, one of them has to ride up over the other. This action will introduce a regional-scale shear to the deforming zone (Figure 12.5b). Which block overrides the other is often

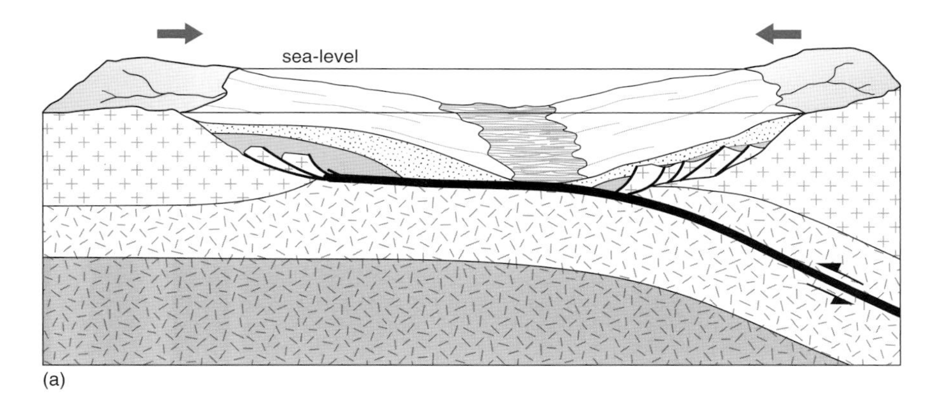

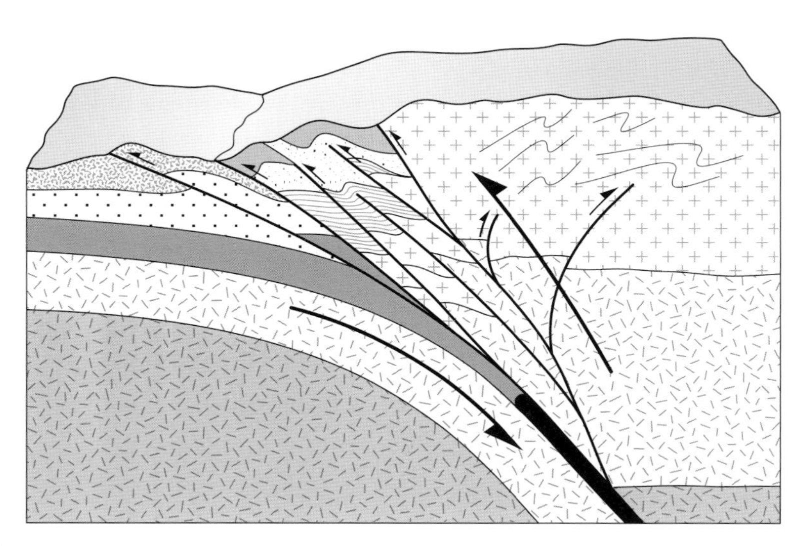

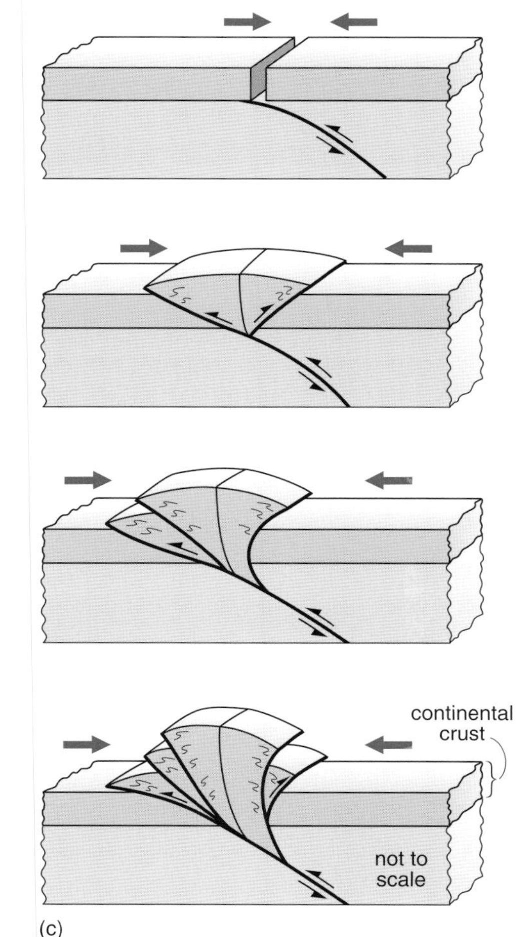

Figure 12.5 Cartoons illustrating the three general points about continental collision made in the text.
(a) Converging continental margins already have an extensional geometry.
(b) Orogenic belts will be asymmetric because one continent must override another. The shear couple (large black half-arrows) will typically be dictated by the direction of subduction.
(c) Within the orogenic belt, deformation will be more intense and slightly older in the inner zone, less intense and slightly younger in the outer zone. Note that the margins of both continents are the first to be involved in deformation. At successive stages, the deformation 'front' moves outwards into the continental block. Sequentially younger thrusts develop in the outer parts of the orogenic belt, carrying deformed rocks in their hangingwalls. The belt as a whole shortens and thickens.

controlled by the polarity of the subduction zone; the continent attached to the subducting oceanic slab tends to dive underneath the approaching continent. Any regional shearing component will control the asymmetry of structures.

Thirdly, we can predict that continental collision will give rise to a wide and varied zone of deformation (Figure 12.5c). Initially, only rocks at the oceanward margin of the continent are deformed. As collision proceeds, deformation spreads outwards and downwards away from the impact zone towards the interiors of both continents. Over time, more and more continental crust becomes incorporated into the orogenic belt. Deformation towards the continental interior will in general be slightly younger than deformation in the impact zone. New thrusts develop in the outermost part of the belt, carrying deformed rocks away from the collision site and towards the continental interiors. Deformation becomes 'frozen' at its maximum extent, when continental collision finally runs out of energy and stops.

Within the orogenic belt, we can distinguish between an *outer* and an *inner* zone. In the outer zone, rocks from the continental interior dominate (e.g. thin layers of sedimentary rocks which rest unconformably on much older metamorphic rocks), and the effects of increased pressure and temperature are less. Deformation is relatively modest, and dominated by thrusts and fault-related fold structures.

In the inner zone, rocks formed at the continental margins are involved in deformation, including thick sedimentary basins and possibly even oceanic crust (as ophiolites) and the subduction zone itself. Deformation is intense and complex; early thrusts and fault-related folds will be overprinted by later ductile folds and schistosities. Because of increased pressure and temperature, medium- to high-grade metamorphic rocks with complex fabrics are to be expected. Crustal melting may occur.

Please note that the outer and inner zones of an orogenic belt are quite different concepts from the paired metamorphic belts you met in Section 8.5.1.

12.3.1 Outer and inner orogenic belts

The outer parts of orogenic belts typically show simple brittle and simple ductile structures in sediments or low-grade metamorphic rocks. Generally, they have deformed over a relatively short period of time, at modest temperatures and pressures. The only high-grade metamorphic rocks present are slices of the 'basement' on which the sedimentary rocks were deposited.

Structures that are typical of the outer parts of many orogens are shown in Figure 12.6. Shortening has been achieved both by folding and thrusting. Thrusts show displacements ranging up to several kilometres, and thrust planes dip towards the inner part of the orogenic belt. Hangingwall displacement is usually away from the site of collision. Folds are commonly asymmetric, open, steeply inclined, and gently plunging. The rocks often carry only a weak, sporadically developed cleavage, and typically only one episode of folding has occurred. The metamorphic record indicates low pressures and temperatures during deformation. Granite plutons are rare or absent.

Many of these features can be seen in Britain's best-preserved outer orogenic belt – the so-called Moine Thrust Belt in the north-west Highlands of Scotland.

❑ Looking at Figure 12.6, when did the folding and thrusting take place?

■ In Tertiary times, because Jurassic, Cretaceous and Tertiary rocks are folded, and Jurassic/Cretaceous rocks are thrust over Tertiary rocks.

The inner parts of orogenic belts typically show complex ductile structures in high-grade metamorphic rocks. Generally, they have suffered repeated phases of

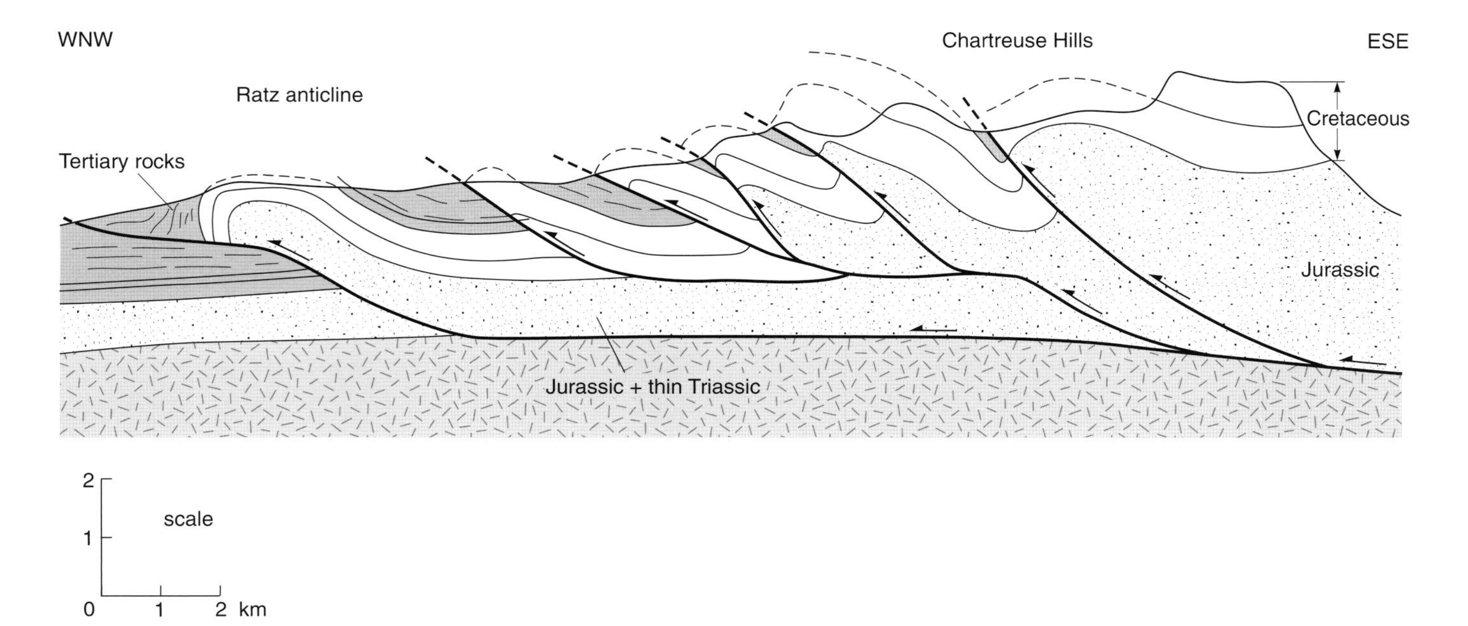

Figure 12.6 A true-scale cross-section through part of the outer zone of the Alps, near Grenoble in France. Triassic and Jurassic rocks are shown stippled, Cretaceous rocks are unornamented, and Tertiary rocks are shown by a grey tone. Heavy lines are thrusts.

progressive deformation, usually over a prolonged period of time and at elevated temperatures and pressures. The continental crust has always been severely shortened and thickened.

Virtually no two inner orogenic belts are alike. Each major orogen that we see around the world tends to have its own distinctive organization of its internal parts. That applies not only to young mountain belts, like the Alps and the Himalayas, but also to the eroded remnants of very old mountain belts, like the Scottish Grampian Highlands. However, several features seem to be common to the inner parts of many orogens.

First, the rocks involved are often much more varied in lithology than those in the outer parts of the orogen. This reflects the complex extensional history that the more oceanward margins of the continents experienced on separation. Significantly, igneous rocks of 'oceanic' affinity often occur here.

Secondly, there are repeated episodes of deformation, usually all recording severe crustal shortening. Often many discrete 'phases' can be recognized, by refolded folds or multiple cleavages. Early folds are commonly asymmetric, tight to isoclinal, flat-lying to recumbent, and usually have a strong axial planar cleavage. Later folds are usually also tight and may be either upright or inclined, and often carry an associated cleavage or schistosity.

Thirdly, the metamorphic record indicates high to extreme pressures, and medium to high temperatures during deformation. Vast granite plutons are extremely common. There is often evidence that highly deformed gneisses have begun to melt (Plate 12.1) to produce some (if not all) of the granite magma.

Fourthly, brittle faults are much less common than ductile folds and cleavages. However, interior zones frequently show large thrust faults that are sometimes associated with immense recumbent folds. They also often contain large, late normal faults that are believed to originate through the 'collapse' effects of gravity on the thickened crust.

The boundary between the outer and inner zones is often not clearly defined and gradational, but equally it may be marked by a clear tectonic line, often an important thrust. The mechanisms by which plate movement causes tectonic structures are still hotly debated amongst Earth scientists. However, the general principles are agreed. Converging plates are driven primarily by gravity, largely the pull of the cold, dense slab of subducting oceanic crust. Continental crust cannot be subducted because it is relatively light. Shear forces generated by the dipping, moving slab act on both continental blocks (Figure 12.5b).

Angled shear produces, in very general terms, horizontal shortening and vertical thickening. Shortening and thickening continue as long as plate movement continues. Throughout deformation, the overthickened continental crust collapses under its own weight, helping to drive thrusting in the outer parts of the orogen, and generating significant internal normal faults.

Some of the world's mountain belts, for example the Urals in Russia, lie towards the centre of continents rather than at their margins. Such mountain chains belong to an earlier cycle of plate tectonics and lie at a join between two once-separate continents. The most important such join close to home lies through Britain and Ireland from the Solway Firth to the Shannon estuary. This line is all that is left of a former ocean that once separated a continental block which included England, Wales and south-east Ireland (and much of Africa) from another including Scotland and north-west Ireland (and part of North America). The ocean that formerly separated these two continents, called the *Iapetus Ocean*, closed about 420 Ma ago (an appropriate name – in Greek mythology, Iapetus was the father of Atlantis, and the Iapetus Ocean can be considered as the forerunner of the Atlantic Ocean). The attendant collision formed a major fold mountain belt, the remnants of which can still be seen in the igneous and folded metamorphic rocks of the Scottish Highlands, and the fold and slate belts of Southern Scotland, the Lake District and North Wales.

12.4 STRIKE–SLIP DEFORMATION AND PLATE TECTONICS

There is one other important type of plate boundary that we have not yet discussed in structural terms.

❑ Which is it?

■ The conservative plate boundary.

Conservative plate boundaries (defined in Section 2.3.4) essentially allow lithospheric plates to move past one another and are dominated by strike–slip movement. In these zones, lateral movement of blocks is the major way in which the crust is deforming; compression and extension play a more minor role. Despite the dominant movement being horizontal, any accommodation structures that develop in strike–slip zones usually also have vertical displacements.

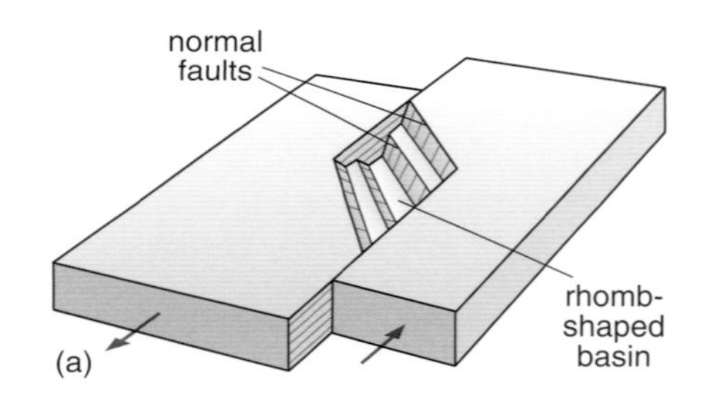

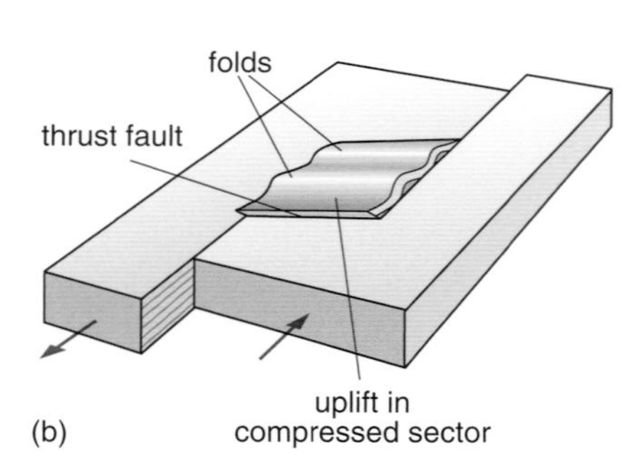

Figure 12.7 Structural features that develop in association with bends in strike–slip systems. (a) Extensional features that develop where strike–slip movement opens up space. (b) Compressional features that develop where strike–slip movement closes down space.

Activity 12.1

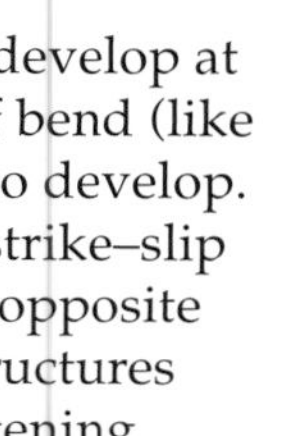

Bends in strike–slip faults cause both crustal extension and vertical movement. Investigate this briefly by using a simple paper model in Activity 12.1.

Natural strike–slip faults have many bends, and the structures that develop at these bends are shown diagrammatically in Figure 12.7. One type of bend (like the first situation in Activity 12.1) allows small sedimentary basins to develop. In these zones, normal faults will form at a high angle to the major strike–slip faults, and open up small, rhomb-shaped basins (Figure 12.7a). The opposite type of bend (like the second situation in Activity 12.1) generates structures that shorten and uplift the rocks near the bend. In these zones, shortening will be accommodated by both thrusting and folding, perhaps with cleavage development (Figure 12.7b). Again, the trace of the structures will be oriented at a high angle to the main strike-slip faults.

Major strike–slip faults typically split into branches or strands which join together further along the length of the fault. Where the main fault splits, or where two strands of the fault overlap, there are generally the same special problems accommodating lateral displacements as those seen at fault bends.

If major strike–slip faults act under a thick sedimentary cover, folds can develop which often show a distinctive pattern above the fault. You can easily get the idea by placing a cloth over two breadboards and then sliding one board sideways past the other; the cloth will ruck up into a set of folds. These folds will be arranged along the underlying fault, but in detail each will be an anticline which plunges in both directions away from its centre, and each will be slightly offset relative to the next (Figure 12.8). This fold arrangement is common in strike–slip zones, but of course the pattern in itself is not especially diagnostic. For example, a set of buried thrust tips could have similar anticlines above them.

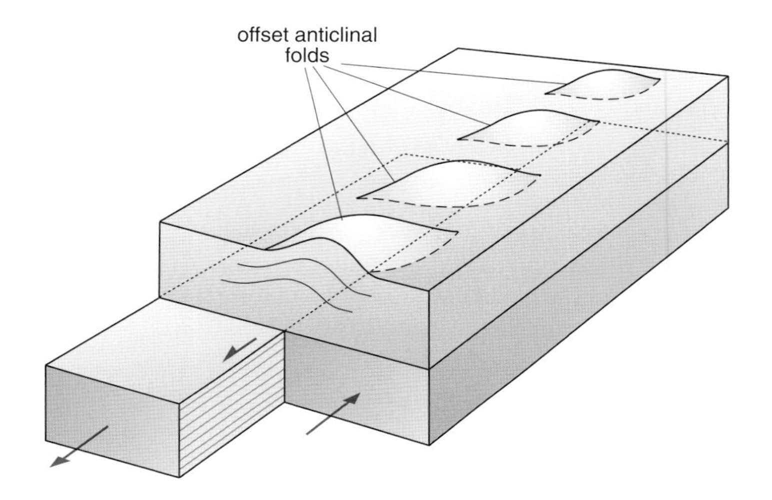

Figure 12.8 Offset anticlines can form in the sedimentary cover above a buried strike–slip fault.

Activity 12.2

In Britain, one of the best examples of a strike–slip fault is the Great Glen Fault in the Scottish Highlands, which runs south-west from Inverness (NH 64 on the Ten Mile Map) through Loch Ness to Loch Linnhe (NM 95). Try Activity 12.2 now, to evaluate the displacement on the Great Glen Fault.

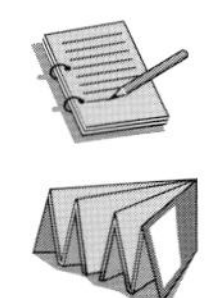

It is clear that strike–slip structures play an important role in deformation, from the metre scale of single faults to the hundreds-of-kilometres scale of conservative plate boundaries. If blocks of crust are to move, they have to be detached from their adjacent crust. Strike–slip structures are one of the two important facilitators of this; the other is large-scale detachments. Consequently, we expect to find strike–slip structures in each of the other tectonic environments, and at all scales. We can also expect these lines, once developed, to be reactivated. The Great Glen Fault is a good example; it is a major strike–slip line that formed within an orogenic zone, yet was later reactivated as part of a basin-forming episode.

However, there are certain distinctive areas within the current plate tectonic cycle where strike–slip deformation dominates in its own right. Considerable lengths of the boundaries of the current plates are conservative in nature – moving past, not towards or away. As we mentioned in Section 2.3.4, the San Andreas Fault zone in Western USA is perhaps the best known continental conservative plate boundary. Indeed, many of our formative ideas on strike–slip tectonics came from this plate boundary. Its sense of displacement is dextral, but the zone itself is not straight, so there are many instances of both sedimentary basins, and zones of fold-thrusts or offset anticlinal folds along its length. It is one of the zones in the world where folding and faulting are actively taking place on the surface and where strain can be measured.

One of the most intriguing aspects of the San Andreas zone is that it is only mid-way through its plate tectonic life. When the Pacific Ocean closes, as it surely will at some time in the future, the structures along the San Andreas will become overprinted by younger, probably more significant structures associated with continental collision.

Activity 12.3

Now complete Activity 12.3, where you are asked to view the final part of the video sequence *Structural geology without tears* on DVD 2.

12.5 Way-up and way forward?

Somewhat earlier in the video, you saw a demonstration of one useful geological 'trick-of-the-trade', when Andrew Bell used sedimentological information from the base and from within a bed to tell where the top of the bed was, and hence work out which 'way-up' the rocks were. He used this information to differentiate anticlines from synclines in awkward recumbent folds.

'Way-up' is one of many practical applications of sedimentology in other branches of Earth Science; applications which encompass topics as diverse as geological time, weathering of mountain ranges, and evolution of life on Earth. In the next and final Block, we turn our attention to the nature of sediments and the sedimentary record.

12.6 Summary of Section 12

- Irrespective of age and global location, certain types of structure repeatedly occur together. The four principal associations of structures which are useful for classification are cratons, sedimentary basins, orogenic belts and strike–slip zones.
- Cratons lie well away from present-day plate boundaries and show no evidence of significant tectonic activity over vast periods of geological time. They typically contain a thin succession of sub-horizontal sedimentary strata, resting unconformably on ancient metamorphic rocks. Over vast areas, these sediments are not significantly folded or faulted.
- Sedimentary basins are areas of continental crust that have thick deposits of sediments deformed by arrays of normal faults. They form by regional surface-parallel extension accompanied by contraction normal to the Earth's surface.
- In many sedimentary basins, an early period of crustal stretching and faulting is succeeded by a later period of thermal collapse without significant faulting.
- Most large sedimentary basins that developed in the past were either intracontinental graben systems, or passive margins that lay between oceanic and continental crust as a result of continental separation.
- Continental collision gives rise to a wide and varied zone of deformation within an orogenic belt. As collision proceeds, deformation spreads outwards away from the impact zone towards the interior of the continent.

Zones are established in which deformation towards the continental interior is generally slightly younger than deformation in the (formerly oceanic) impact zone. Deformation becomes 'frozen' at its maximum extent when continental collision ends.

- The outer parts of orogenic belts typically show simple structures in sediments or low-grade metamorphic rocks. Shortening is achieved both by folding and displacement on thrusts. Thrusts dip towards the inner part of the orogenic belt, and commonly show hangingwall movement away from the site of collision. Folds are commonly asymmetric, open, steeply inclined and gently plunging. The rocks themselves usually carry only a weak cleavage. Often only one episode of folding has occurred.
- The inner parts of orogenic belts typically show complex ductile structures in high-grade metamorphic rocks. They show repeated episodes of deformation. Early folds are commonly asymmetric, tight to isoclinal, flat-lying to recumbent, and usually have a strong axial planar cleavage. Later folds are usually also tight and may be either upright or inclined, and often carry an associated cleavage. Their metamorphic record indicates high pressures and temperatures during deformation. Vast granite plutons are extremely common.
- Strike–slip structures play an important role in deformation, from metre scale to the scale of conservative plate boundaries. In some respects, conservative boundaries represent incomplete stages of the tectonic part of the plate tectonic cycle.

12.7 Objectives for Section 12

Now you have completed this Section, you should be able to:

12.1 Describe the main tectonic associations found across the world.

12.2 List the main structural features by which you would recognize sedimentary basins, the outer and inner parts of orogenic belts, and strike–slip zones.

12.3 Outline how each tectonic association fits into the plate tectonic cycle.

Now try the following questions to test your understanding of Section 12.

Question 12.1 Describe the changes in sedimentary thickness you would expect to see moving from a cratonic continental interior onto and across a passive continental margin and onto the deep ocean floor. (*A few sentences*)

Question 12.2 Describe the structures you would expect to see moving from a cratonic continental interior onto and across a passive continental margin. (*A few sentences*)

Question 12.3 From Figure 12.6, what is the % extension in this part of the outer Alpine orogenic belt: –10%, –30% or –50%? First take a guess, to see how good your 'feel' for the amount of deformation is, then make a more accurate calculation to check how good your guess was. (*Tip:* you need to know the length of any one bed before and after deformation. We recommend you use the bed in the middle of the Cretaceous rocks.)

ANSWERS TO QUESTIONS

Question 2.1

One possible explanation is that rocks get denser at greater depths. This makes sense because the weight of the overlying material must cause pressure to increase as depth increases, and you would expect rock at depth to be compressed to greater densities. (*Comment*: this effect is described as 'self-compression'.) The Earth's interior would therefore be denser than the Earth's surface even if the interior and surface were made of material of identical composition. An alternative explanation is that the interior consists of something naturally denser than the familiar silicate rock types, maybe something that is not rock at all.

Question 2.2

(a) The SiO_2 content of oceanic crust falls into the range of SiO_2 contents for mafic igneous rocks given in Table 6.4 of Block 2. Furthermore, its overall composition is a close match to the gabbro shown in Table 6.3 of Block 2.

(b) The SiO_2 content of continental crust falls into the range of SiO_2 contents for intermediate igneous rocks given in Table 6.4 of Block 2. Note that the approximately equal proportions of Na and K in Table 2.2 are consistent with an intermediate rock type.

Question 2.3

The most obvious reason is that oceanic crust is thinner than continental crust, so it 'floats' lower. The second reason is that it is denser than continental crust, so it would be less buoyant than continental crust of the same thickness. There is actually a third reason, too, which is that ocean basins are flooded by seawater, which will itself tend to depress the ocean floor so the weight of this water has to be taken into account when calculating the mass per unit area of each column (we can, however, ignore the effect of the atmosphere, because the differences between atmospheric mass per unit area overlying each column are negligible).

Question 2.4

The calculation is very simple. You must divide the distance by the speed, making sure that you use the same unit of distance. 80 km is 8×10^4 m, and 1 cm yr^{-1} is 1×10^{-2} m yr^{-1}. The time to move this distance at this speed is

$$\frac{8 \times 10^4 \text{ m}}{1 \times 10^{-2} \text{ m yr}^{-1}} = 8 \times 10^6 \text{ years, or 8 million years.}$$

Question 2.5

(a) The way to do this is to calculate the heat produced by each isotope per kg of rock and add these values together. Remember that Table 2.4 quotes concentrations in parts per million, so you must include a factor of 10^{-6}. For example, basalt contains 0.4 ppm of ^{40}K, so 1 kg of basalt contains 0.4×10^{-6} kg of ^{40}K. We know from Table 2.3 that the heat produced per kg of ^{40}K would be 2.8×10^{-5} W, so the heat production from ^{40}K in basalt is thus $0.4 \times 10^{-6} \times 2.8 \times 10^{-5}$ W kg^{-1}.

We can conveniently combine the powers of ten and write this as $(0.4 \times 2.8) \times 10^{-11}$ W kg^{-1}. Doing the same for the other three heat-producing isotopes and adding the values together, the combined expression for heat production from ^{40}K + ^{232}Th + ^{235}U + ^{238}U in basalt is:

$$(0.4 \times 2.8 + 2 \times 2.6 + 0.003 \times 56 + 0.497 \times 9.6) \times 10^{-11} \text{ W kg}^{-1}$$
$$= (1.12 + 5.2 + 0.17 + 4.8) \times 10^{-11} \text{ W kg}^{-1}$$
$$= 11.3 \times 10^{-11} \text{ W} = 1.13 \times 10^{-10} \text{ W kg}^{-1}$$

which when expressed to two significant figures is the value in Table 2.4.

The equivalent working for peridotite is:

$$(0.04 \times 2.8 + 0.06 \times 2.6 + 0.00014 \times 56 + 0.02 \times 9.6) \times 10^{-11} \text{ W kg}^{-1}$$
$$= (0.112 + 0.156 + 0.008 + 0.192) \times 10^{-11} \text{ W kg}^{-1}$$
$$= 0.468 \times 10^{-11} \text{ W kg}^{-1} \approx 0.047 \times 10^{-10} \text{ W kg}^{-1}$$

which is the value given in Table 2.4.

(b) If you were able to verify the values of heat production in basalt and peridotite in part (a), then you should have been able to calculate the heat production in granite using the same method. The expression is:

$$(4 \times 2.8 + 20 \times 2.6 + 0.03 \times 56 + 3.97 \times 9.6) \times 10^{-11} \text{ W kg}^{-1}$$
$$= (11.2 + 52 + 1.68 + 38.1) \times 10^{-11} \text{ W kg}^{-1}$$
$$= 103 \times 10^{-11} \text{ W kg}^{-1} = 10.3 \times 10^{-10} \text{ W kg}^{-1}$$

The third significant figure is not justifiable, because the elemental abundances you used were quoted to only one significant figure, so it would be preferable to enter a value of 10 rather than 10.3 into the empty cell in Table 2.5.

Question 2.6

(a)(i) Mass/volume = density. The fact that the Earth's average density is so much greater than the density of material near the surface tells us that there must be something unusually dense (which we now know to be the core) deep in the interior.

(ii) Seismic studies reveal the compositional layers, in particular the boundaries between inner and outer core, between outer core and mantle, and the Moho between crust and mantle. There are also jumps in seismic speeds attributed to phase transitions, such as the one between the upper and lower mantle.

(iii) The Earth's magnetic field indicates a fluid and electrically conducting core.

(b) The lines of evidence are mutually consistent, but the most important is the seismic evidence, because it enables the depths of the boundaries between each layer to be measured.

Question 2.7

(a) Removal of material from the top of the crust means that the crust has become thinner. In order to maintain isostatic equilibrium, the base of the crust will rise. (*Comment*: as the base rises, it must push the whole column up with it, which will tend to encourage even more erosion at the top.)

(b) Deposition of sediment on top of column D will add an extra load, so that whole column will be isostatically depressed.

(c) Flow takes place in the asthenosphere, which is below the bottom of this diagram. The mantle within the diagram is part of the lithosphere and will bend (flex) to allow the crust to rise and fall, but is too strong to flow.

Question 2.8

These are: (i) isostasy, as explored in Question 2.7c; (ii) plate tectonics, which can only happen because there is a weak layer below the lithosphere; (iii) the slowing down of seismic waves in the low velocity zone, indicative of a few percent of melt between grains and which is therefore particularly weak.

Question 2.9

False. There is a constructive plate boundary (the Mid-Atlantic Ridge) running down the centre of the Atlantic (shown in Figure 2.5). Therefore, North America and the western part of the North

Atlantic lie on a different plate to Europe and the eastern part of the North Atlantic. Similarly, South America and the western part of the south Atlantic lie on a different plate to Africa and the eastern part of the South Atlantic.

Question 2.10

Heat-producing elements are more abundant in continental than in oceanic crust, so the setting with the thickest continental crust is where most heat will be generated. This is (c), where the crust has been thickened by collision. (*Comment*: As you will see in Section 7.5, the granite intrusions in many mountain belts were formed by melting caused by this enhanced heat production.)

Question 3.1

Only the oceanic crust that is added to the Nazca Plate is relevant. This will be transported eastwards towards South America, at a closure rate of a few cm yr^{-1}. When it reaches the trench, it will be subducted. As it goes down, it will begin to melt. This melt will rise upwards, and some of the melt will form intrusive rocks within the South American crust and the rest will be erupted at volcanoes (which tend to be about 100 km above the descending plate).

Question 4.1

(a) The positions of the four points specified are plotted on Figure A4.1. Points (i) and (iii) lie between the solidus and the liquidus, so at these conditions mafic material would consist of a mixture of solid and liquid. Point (ii) lies to the left of the solidus and represents entirely solid conditions. Point (iv) lies to the right of the liquidus, and represents entirely liquid conditions.

(b) At 0.1 MPa, the solidus lies at a temperature of 1080 °C, so melting begins at this temperature. At the same pressure, the liquidus lies at 1200 °C, so the sample would be completely molten at this temperature.

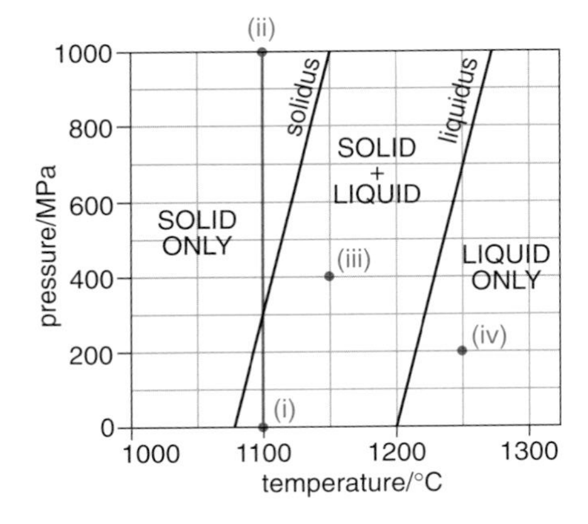

Figure A4.1 Annotated version of Figure 4.1. Points (i) to (iv) are referred to in the answer to Question 4.1. The line linking points (i) and (ii) is referred to in the answer to Question 4.2.

Question 4.2

Initially, the basalt will be entirely solid (the *P–T* conditions correspond to point (ii) on Figure A4.1). The *P–T* conditions experienced by the sample will vary along the straight line drawn from (ii) to (i) on Figure A4.1. Point (i) corresponds to the final state, which is a mixture of crystals and liquid. The first liquid will form under the conditions represented by the point where our line crosses the solidus, which is when *P* has decreased to about 300 MPa. The sample will be entirely solid until *P* has fallen to this value, and as *P* continues to decline the proportion of liquid will increase.

Question 4.3

The *P–T* conditions are those of point (ii) in Question 4.1a. As we saw, under anhydrous conditions this corresponds to completely solid basalt. However, if conditions become water-saturated, then the water-saturated phase boundaries apply. The water-saturated liquidus lies to the left of this point, so our sample would be completely molten (and would therefore be water-saturated mafic magma).

Question 4.4

The first melt to form during partial melting of a mafic rock will tend to be intermediate in composition, because most of its constituents would come from pyroxene and the least calcium-rich plagioclase (Section 4.2). However, even if the first tiny fraction melt were felsic, this would be particularly viscous and would not easily flow. As partial melting proceeded, the melt would become progressively less felsic, and the average composition of the melt would certainly be intermediate before enough melt (>5%) had collected to be able to rise freely.

Question 4.5

(a) At 30 °C km^{-1}, temperature at a depth of 30 km would be higher than the surface temperature by 30 °C km^{-1} × 30 km = 900 °C. Surface temperature is 10 °C, so the temperature at 30 km is 910 °C.

(b) The point representing a depth of 30 km and a temperature of 910 °C is labelled A on Figure A4.2. This is to the left of the anhydrous solidus, showing that under these conditions felsic material would be entirely solid.

(c) The water-saturated phase boundaries are now applicable. Point A lies to the right of the water-saturated liquidus, so the material will be completely molten.

(d) The line representing cooling at 5 °C km^{-1} is plotted on Figure A4.2 (at 5 °C km^{-1}, cooling during a rise of 30 km would be 5 °C km^{-1} × 30 km = 150 °C). Our magma would be completely liquid until the depth where this cooling line crosses the liquidus (point B on Figure A4.2), which in this example is at 3.8 km depth (100 MPa). The magma will begin to crystallize at this depth, and become progressively more crystalline as it continues to rise

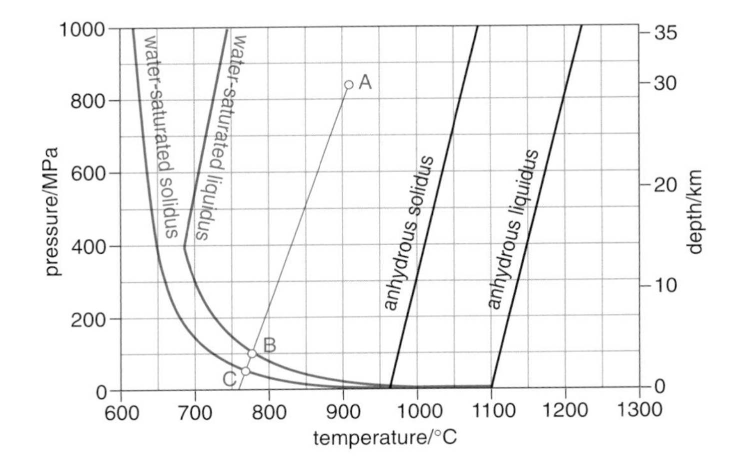

Figure A4.2 Annotated version of Figure 4.3, to accompany answer to Question 4.5.

(unless fractional crystallization allowed the melt to be separated from the crystals, which we will ignore). The magma would be completely crystallized by the time its cooling path intersects the water-saturated solidus, at about 2.3 km (point C on Figure A4.2), and so it would cease to rise at this depth, if not slightly deeper. (*Comment*: This answer ignores the role of volatiles, which will tend to exsolve as the pressure declines. We will return to that issue in Section 6.)

Question 5.1

(a) The low % of quartz puts this rock somewhere on the lower row of Table 5.1. To place it within this row, we have to calculate the ratio $\frac{\text{plagioclase}}{\text{plagioclase + potassium feldspar}}$, which in this example is $\frac{40}{100} = 40\%$. This places the rock in the monzonite box, monzonite being the correct name for a coarse-grained rock of this composition.

(b) The high percentage of quartz puts this rock somewhere on the upper row of Table 5.1. The feldspar ratio, calculated as before, is $\frac{70}{100} = 70\%$. This places the rock in the granodiorite box, and microgranodiorite is the correct name for a medium-grained rock of this composition.

(c) The low % of quartz puts this rock somewhere on the lower row of Table 5.1. The feldspar ratio, calculated as before, is $\frac{85}{100} = 85\%$. This places the rock in the 'diorite or gabbro' box. To decide whether it is dioritic or gabbroic, we have to take account of the composition of the plagioclase, which we are told is 70% albite. As this is more than 50%, the rock is dioritic rather than gabbroic. If it were coarse-grained, it would be a diorite, but in this case we are told it is fine-grained so it should be called an andesite.

Question 5.2

(a) During emplacement of cone sheets, the block labelled A is uplifted. This implies that the magma chamber is inflating, causing forceful injection of cone sheets through the updoming roof.

(b) During emplacement of a ring dyke, block A subsides along the associated fracture. Because of the outward dip of the fracture, this movement widens the fracture and magma must be displaced upwards to occupy the space. This is a more passive mode of emplacement than for cone sheets.

Question 5.3

(a) Thin section (a) has a clearly visible groundmass of crystals (in fact mostly plagioclase and pyroxene) about 0.1 mm or less in size. Fine grain size is classified as < 0.25 mm, so this rock is fine-grained. Most of thin section (b) has no discernible crystals, and consists of a dark (in fact brown) glass.

(b) The glassy groundmass of thin section (b) shows that it comes from closer to the margin. In fact (b) is from within 1 cm of the margin, whereas (a) comes from several cm further in.

(c) The relatively large plagioclase feldspar crystals must have begun to grow before the magma was emplaced (like the pyroxenes in the Whin Sill chilled margin in Plate 5.3). In (a), it is surrounded by fine grained crystals that grew in the rapidly cooling magma as it lost heat to the country rocks, and in (b) it is enclosed in glass resulting from the even more rapid chilling right against the contact with the country rock. Because they are so much bigger than the groundmass, these crystals can be described as phenocrysts (Block 2 Section 6.1) or, because they are rather small, as microphenocrysts.

Question 5.4

(a) This is a basaltic dyke. Its composition is given by the number 35 and that it is a dyke can be inferred from its narrow, near linear outcrop discordant with the sedimentary units it crosses. The youngest of these is 85 (Lower Permian), so we can tell that the dyke must be younger than the lower Permian. (In fact this is one of the radial dykes from Mull, and is lower Tertiary in age.)

(b) This is a granite pluton, and is discordant with the country rock, cutting across the boundary between 70–1 (Ashgill and Caradoc; Ordovician) and 72 (Llandovery; Silurian). We can tell therefore that it must be post-Llandovery in age. (In fact it is of Devonian age, similar to the Shap granite.)

Question 5.5

With only 10% quartz it lies in the bottom row of Table 5.1. With four times as much potassium feldspar as plagioclase feldspar, the $\frac{\text{plagioclase}}{\text{plagioclase + potassium feldspar}}$ ratio is $\frac{1}{1+4}$ or $\frac{1}{5}$, which is 20%. This puts the rock in the syenite box, but as its grain size is fine we should call it a trachyte.

Question 5.6

(a) The combined surface area of these granites exceeds 100 km^2, so they could legitimately be described as a batholith. The Starav Granite has been intruded into the centre of the Cruachan Granite, so they form a concentrically intruded pluton (similar to the Rogart intrusion you studied in Activity 5.1). The ring-like outcrop of granite at the northern end of the Starav Granite is much too wide to be a cone sheet. As far as we can tell from this map, it could be either a ring fracture below a pluton or a ring dyke above a pluton (in fact it is the latter, and is associated with volcanic rocks in the Glen Coe area). The dykes, although clearly related to the plutonic complex, are not radial to it, and so should be described simply as a dyke swarm.

(b) The dykes cut the Cruachan Granite, so they must be younger than this. However, they themselves are truncated by the Starav Granite, which must therefore be younger still. (*Comment*: It could be argued that the dykes were fed from the Starav Granite, and therefore be of the same age, but this possibility can be discounted because their compositions are different.) The sequence of intrusion is thus: first the Cruachan Granite, then the dyke swarm, and finally the Starav Granite. (*Comment*: The Cruachan and Starav Granites have been dated radiometrically as 401 Ma and 396 Ma respectively. However, the uncertainties in these measurements are ± 6 Ma and ± 12 Ma respectively, so without the field (map) evidence the sequence of intrusion could not be proven.)

Question 6.1

Note: You have already examined a thin section of a basalt lava (TS I) in Activity 6.1 of Block 2 and it might be a good idea for you to look briefly at this again. Irrespective of the characteristics of that particular specimen, we would expect lava flows to be fine-grained because of the rapid cooling they must experience. Particularly in thin flows, we might find glass between the fine crystals. The surfaces of some flows, notably pahoehoe basalts, are glassy because of rapid chilling. Rhyolites are noted for being glassy throughout in many instances. Really thick flows might cool slowly enough in their centres to result in a medium grain size. Any flow fed from a magma chamber in which crystallization had been taking place could contain phenocrysts (e.g., the andesite, RS 1 in your Home Kit), irrespective of the grain size of its groundmass.

Question 6.2

In case (a), the cooling line (C–D on Figure A6.1) would reach the surface at 860 °C (it would be the cooling line as drawn in Figure A6.2 displaced to the right by 100 °C); and in case (b), the cooling line (A–E on Figure A4.1) would reach the surface at 850 °C (it would be a steeper cooling line than in Figure A4.2, but its upper end would still begin at point A). In both cases, the cooling line reaches the liquidus at a depth of about 1.5 km (which is where magma would begin to crystallize) and the solidus (where it would be completely solid) at a depth of less than 0.5 km.

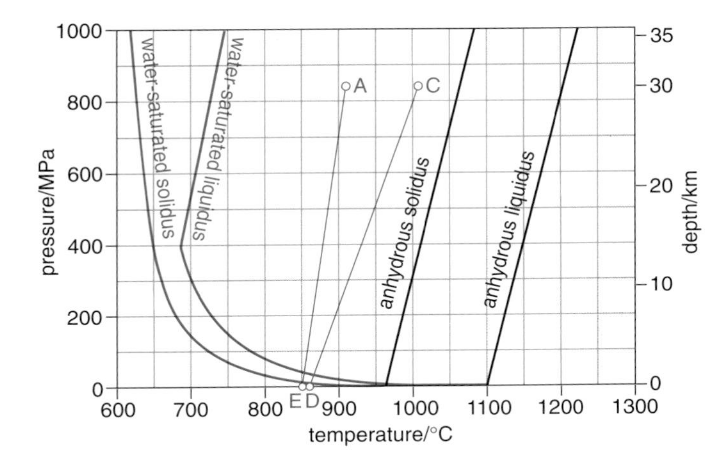

Figure A6.1 Annotated version of Figure 4.3, to accompany answer to Question 6.2.

Question 6.3

The contours of tephra thickness are displaced towards the south-south-east, indicating that the umbrella cloud was dispersed in this direction. The wind must therefore have been blowing from the north-north-west.

Question 6.4

(a) The initial point (A on Figure A6.2) plots in the stability field for a convecting column, so there would be a plinian eruption. However, the point representing a 200 m radius vent at 300 m s^{-1} exit velocity (B on Figure A6.2) plots in outside the field for a stable column, and the column would collapse, generating a pyroclastic flow.

(b) Just as in (a), the new point (C on Figure A6.2) would plot in the collapsing column field, so this change in eruption conditions would also lead to column collapse.

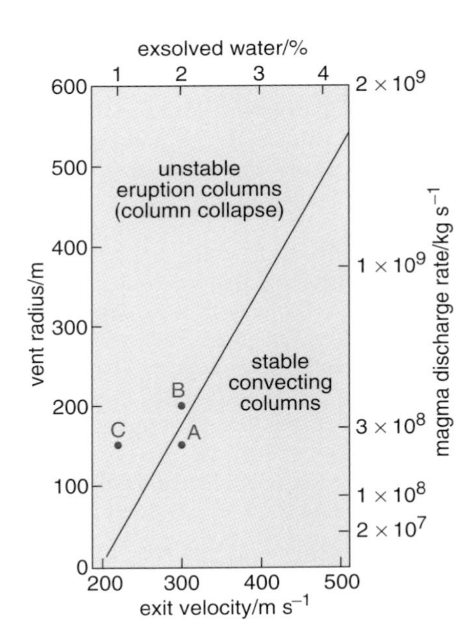

Figure A6.2 Annotated version of Figure 6.15, to accompany answer to Question 6.4.

Question 6.5

Airfall will tend to lie where it falls, and because topography can have no influence on the fall process the thickness of an airfall deposit will be independent of topography but will reflect distance from the vent and direction of the wind (Figure A6.3a). However, because pyroclastic flows travel along the ground, we expect them to travel primarily down slopes and to be channelled along valleys. Flow deposits should therefore be thicker in the flat parts of valley bottoms (Figure A6.3b).

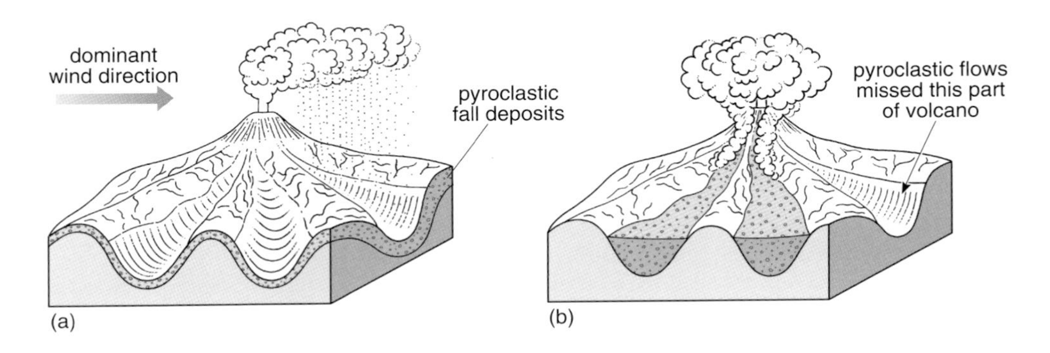

Figure A6.3 (a) Airfall deposits are distributed around a volcano independently of topography, but controlled by wind direction. (b) Pyroclastic flow deposits are largely confined to valleys. (To accompany the answer to Question 6.5.)

Question 6.6

The correct answers are shown below:

	Low volatile content	High volatile content
(a) mafic magma	fluid lavas	fire fountaining
(b) felsic magma	viscous domes	pyroclastic flows and airfall

If you got any of these wrong, you should re-study Sections 6.1 and 6.3.

Question 6.7

There could be a major caldera-forming event such as the one that created Crater Lake (Figure 6.24), triggered by eruption of a large volume of generally felsic magma from the magma chamber. Shallow-level intrusion (as at Mount St Helens, Fig. 6.18) or weakening by a fissure system could cause one side of the volcano to bulge and then to collapse. A fissure system could also channel eruptions to the surface along a volcanic rift, so that the volcano developed an elongated shape (like Mauna Loa and Kilauea, Hawaii, Figure 6.25, although that example is a shield volcano rather than a stratocone).

Question 6.8

An ignimbrite usually contains pumice clasts. These tend to be absent in block-and-ash flow deposits, which are rich in denser (lithic) clasts. The pumice in an ignimbrite is particularly easy to distinguish if it has been squashed into fiamme. Both deposits have an ashy matrix, which is sometimes welded in an ignimbrite. Care would need to be taken if the exposure is incomplete, because pumice clasts are rare near the base of reverse-graded ignimbrites. Either deposit could be associated with an airfall deposit, so this would not be useful diagnostic evidence.

Question 6.9

A completed Table is shown below:

	Sill	Lava flow
chilled margin at top	✓	(✓)
chilled margin at bottom	✓	(✓)
columnar joints	✓	(✓)
concordant base with local discordance	✓	(✓)
concordant top with local discordance	✓	×
rubbly top	×	✓
rubbly bottom	×	✓
baking of overlying rock immediately above the contact	✓	×
baking of underlying rock immediately below the contact	✓	✓

Proper chilled margins occur in sills but not lava flows. However, the glassy top of some kinds of pahoehoe (especially of the sheet-flow variety) might be mistaken for a chilled margin, which is why we have put a tick in brackets in the lava column. Similarly, the base of a lava flow might be chilled against the underlying rock. Therefore chilled margins are not infallible guides. Columnar joints can occur in thick lava flows as well as in sills. Sills are generally concordant, but can be locally discordant at top or base. The base of a lava flow is discordant if there has been erosion prior to its eruption, but otherwise concordant. However, deposits laid on top of the flow will be concordant *everywhere*, with no local discordances. Rubbly tops and bottoms are characteristic of a'a flows and blocky flows, and offer the most reliable means in this Table of distinguishing a lava flow from a sill. Sills cause baking of the rock into which they are intruded (Section 5.2) and similarly we would expect the heat from a lava to bake the immediately underlying rock. Material deposited on top of a lava flow cannot be baked, though, because the flow would be cold by then.

(*Comment*: A further criterion, not listed in the Table, is weathering. The top of a lava flow may show the effects of long-term exposure prior to burial as seen in Video Band 4, whereas this cannot occur in a sill.)

Question 7.1

(a) At 2000 Mpa, the 600 °C isotherm passes through the base of the subducting crust in the slab. The top of the crust is hotter, in fact nearly 700 °C based on interpolating between the 600 °C and 800 °C isotherms.

(b) At 2000 MPa, the hydrous solidus for basalt lies at about 630 °C. This means that any mafic rock hotter than 630 °C at this pressure will be partially molten. In Figure 7.5, most of the oceanic crust at this depth meets that condition, so partial melting will occur throughout all but the lowest (coldest) part of the subducting crust.

Question 7.2

The answer can be extracted from Figure 7.6 and the associated text. One source is hydration of the lower continental crust by water driven off the slab. Depending on the composition of the lower crust, this magma could be either felsic (which is what we want) or intermediate in composition. If intermediate, it would need to undergo fractional crystallization prior to eruption to remove the crystals forming at highest temperature, thereby increasing the silica content of the remaining melt until it became felsic. Another source is partial melting of the crustal part of the slab, yielding an intermediate magma, which would need to undergo fractional crystallization as above before eruption. (*Comment*: Felsic magma can also arise from extreme fractional crystallization of mafic magma from other sources indicated in Figure 7.6, but the total volume produced in this way is probably less than for either of the two main sources.)

Question 7.3

At the location indicated, the temperature is approximately 1100 °C. The mantle here will be hydrated because of water driven off from the slab, so the relevant phase boundary on Figure 7.2 is the hydrous solidus for peridotite. The point at 1100 °C and 3000 MPa falls to the right of this solidus, so the mantle would be undergoing partial melting. (*Comment*: However, if the mantle were anhydrous there would be no melting, because this point lies to the left of the anhydrous solidus.)

Question 8.1

(a) The phases tremolite and calcite and quartz are all solids (minerals) and thus this assemblage is *more* ordered, and so has a *lower* entropy than the assemblage diopside + H_2O + CO_2 which contains two gas phases.

(b) The higher entropy assemblage, in this case diopside + H_2O + CO_2, is stable at higher grades of metamorphism.

Question 8.2

(a) At 300 MPa and 700 °C, a sample would lie in the sillimanite stability field (see A, Figure A8.1), thus andalusite would not be stable.

(b) At 800 MPa and 500 °C, a sample would lie in the kyanite field (B, Figure A8.1), so kyanite would be the stable phase.

(c) Kyanite and sillimanite only coexist in equilibrium on the phase boundary between their respective stability fields. At 700 MPa, the temperature on that phase boundary is *c.* 660 °C (C, Figure A8.1).

(d) Kyanite is metastable, since at 400 °C and at atmospheric pressure andalusite must be the stable phase (D, Figure A8.1).

Question 8.3

(a) Basalts contain feldspar, and although its composition may change during metamorphism, feldspar is often present in metamorphosed basalts. At high grades of metamorphism, garnet may form from the elements Fe, Mg, Ca and Al. However, there is insufficient K to form muscovite, and absence of carbon (in the form of carbonate ions) prevents the formation of calcite.

(b) Limestones are rich in Ca and CO_2, and thus calcite ($CaCO_3$) will be present after metamorphism. Conversely, metamorphosed limestones contain little or no aluminium, iron, magnesium or potassium and hence cannot form garnet (which requires Fe, Mg and Al) or muscovite (which requires K and Al) or feldspars (which need Al).

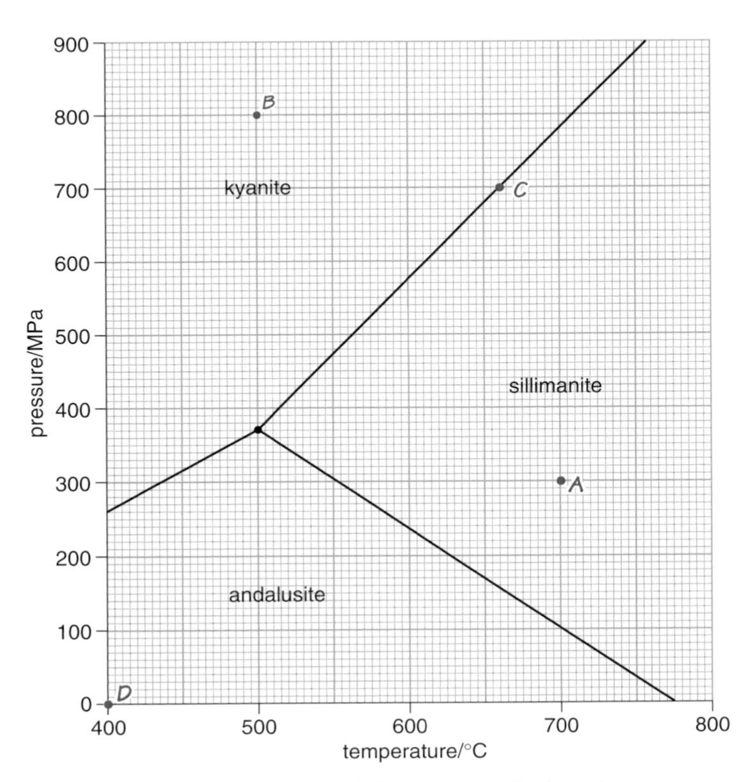

Figure A8.1 Phase diagram of the polymorphs kyanite, andalusite and sillimanite with points A, B, C, D marked for answer to Question 8.2.

(c) Garnet, muscovite and feldspar occur in metamorphosed mudstones which contain Fe, Mg, Ca and Al (see Block 2 Table 7.2). Absence of carbonate prevents calcite from forming.

Question 8.4

(a) Since andalusite rather than kyanite reacts to form sillimanite in the metamorphic aureole in Figure 8.6, the pressure must be less than about 370 MPa. Andalusite is not stable at higher pressures (Figure 8.2).

(b) At the outer margin of the sillimanite zone in Figure 8.6, andalusite presumably reacts to form sillimanite. At a pressure of 200 MPa, this takes place at 625 °C according to the phase diagram in Figure 8.2.

Question 8.5

The reaction chlorite + quartz $\rightleftharpoons$ garnet + H_2O is temperature-dependent and almost independent of pressure at pressures greater than about 500 MPa; that is, the phase boundary is almost vertical on the phase diagram of pressure against temperature (Figure 8.8). At pressures of >370 MPa, garnet forms from quartz and chlorite at temperatures of 552 to 570 °C.

Question 8.6

(a) Only two isograds can be drawn on Figure 8.13; a garnet isograd between the biotite and the garnet zones, and a kyanite isograd between the garnet and kyanite zones (Figure A8.2).

(b) The isograds cut across the folds and so the metamorphism is presumably later than the folding.

(c) The isograds are not concentric around the intrusion, and the latter occurs in an area of relatively low metamorphic grade (the biotite zone), and so the metamorphism is not related directly to the intrusion. It is not therefore an example of contact metamorphism and so presumably results from regional metamorphism.

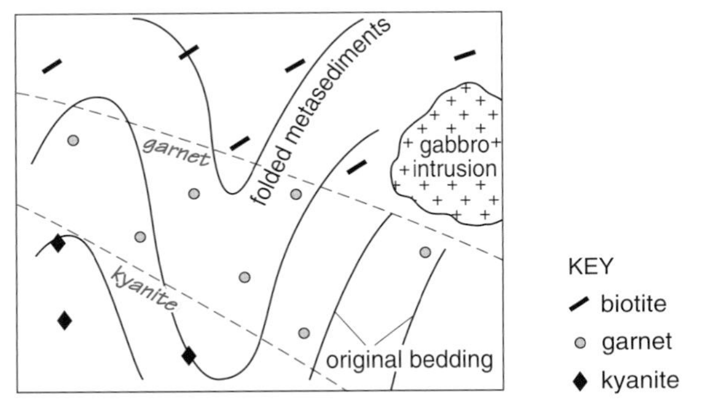

Figure A8.2 Geological sketch map for answer to Question 8.6.

Question 8.7

Point A is in a mafic igneous rock within the sillimanite zone, and so should contain amphibole and plagioclase feldspar and garnet (Table 8.1). Point B is in a carbonate in the kyanite zone, and so should contain garnet, anorthite and amphibole. Point C is in a carbonate in the garnet zone, and so should contain garnet, epidote and amphibole. Point D is in a carbonate in the staurolite zone, and so should contain garnet, anorthite and amphibole.

Question 8.8

At A, $T = 620$ °C, $P = 400$ MPa. This indicates a high geothermal gradient and, according to Figure 8.11, amphibolite-facies conditions. (Near A, metamorphic facies would increase rapidly with depth from the surface from greenschist to amphibolite and ultimately granulite facies.) At B, $T = 290$ °C, $P = 1600$ MPa corresponding to blueschist facies on Figure 8.11 and indicating a very low geothermal gradient.

Question 8.9

Water as either a liquid or a vapour is less ordered and thus has a higher entropy than the other phases, which are all (solid) minerals. The assemblage staurolite and quartz therefore has the lower entropy, so the assemblage garnet and sillimanite + H_2O, which has the higher entropy, must be stable at higher grades of metamorphism. Conventionally, an isograd is named after the index mineral on the side of *higher* metamorphic grade. Thus, this reaction might have taken place at the sillimanite isograd.

Question 8.10

The presence of staurolite indicates that the temperature was greater than 580 °C. As kyanite is present at what appears to be moderately high temperatures (>580 °C), the pressure must have been greater than 530 MPa (Figure A8.3). The fact that these pelitic rocks have not melted to form granites implies the temperature is less than 625 °C at these pressures. Thus, during metamorphism the temperature was between 580 °C and 625 °C at pressures greater than 530 MPa (see red shaded area on Figure A8.3). This assemblage is from the central Alps in Switzerland, and other evidence suggests that the pressures were 600–700 MPa.

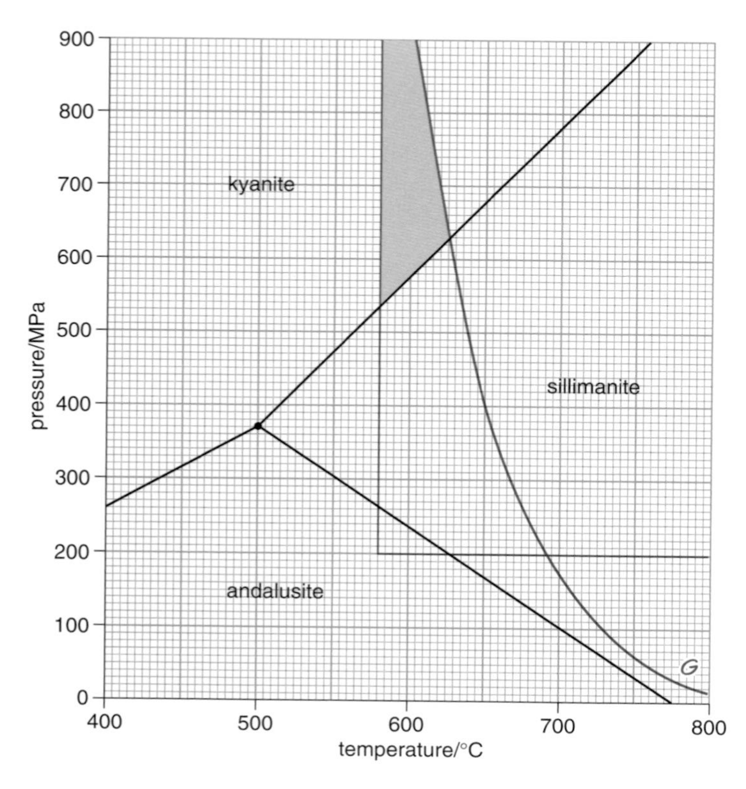

Figure A8.3 Phase diagram for answer to Question 8.10. G = water-saturated granite solidus.

Question 8.11

At 800 MPa and 200 °C, a rock belongs to the blueschist facies (Figure 8.11). 800 MPa corresponds to a depth of $\frac{800\,\text{MPa}}{30\,\text{MPa km}^{-1}}$ = 26.7 km, so (assuming a surface temperature of 0 °C) the geothermal gradient is $\frac{200\,°\text{C}}{26.7\,\text{km}} = 7.5\,°\text{C km}^{-1}$. This represents relatively high pressure–low temperature metamorphism of the kind typically found associated with subduction at destructive plate boundaries.

At 200 MPa and 550 °C, a rock plots in the hornfels facies near the transition into the amphibolite facies (Figure 8.11). This is relatively high temperature–low pressure metamorphism. 200 MPa corresponds to a depth of $\frac{200\,\text{MPa}}{30\,\text{MPa km}^{-1}} = 6.7$ km so (assuming a surface temperature of 0 °C) the geothermal gradient is $\frac{550\,°\text{C}}{6.7\,\text{km}} = 82\,°\text{C km}^{-1}$, which is much higher than that predicted by models of, for example, crustal thickening (curve 2, Figure 8.14). It is most likely to be associated directly with magmatic activity in which hot magma has significantly increased the temperature at relatively shallow depths. At present, most magmas, other than those formed at constructive boundaries, are generated above subduction zones in island arcs, or above destructive continental plate boundaries such as the South American Andes.

Question 9.1

Using the information that the fenceposts are *c.* 1 m long and the road is *c.* 5 m from white line to kerb, the answer is as follows. The wavelength, from anticlinal crest on the extreme left of the photo to the prominent anticlinal crest in the centre right, is approximately 15–20 m, and the amplitude (half the height from anticline crest to syncline crest in the same bed) is about 3–5 m.

Question 9.2

Both are dip–slip faults in which the hangingwall has moved up relative to the footwall. However, thrusts have low fault-plane dips, less than 45° and typically less than about 20°, whereas reverse faults have steeper dips, over 45°.

Question 9.3

Normal faults have relatively steep fault planes, typically dipping at 60° or so, with mostly dip–slip movement in which the hangingwall is displaced down-dip. *Thrust faults* have relatively flat-lying fault planes, typically dipping 0–20°, with mostly dip–slip movement in which the hangingwall is displaced up-dip. *Strike–slip faults* have very steep fault planes with almost all horizontal and relatively little vertical movement. Displacement can be either to the right (dextral) or to the left (sinistral).

Question 9.4

Joints are discrete fracture planes through rock that separate blocks of undeformed rock. This is consistent with brittle deformation, but not with ductile deformation since that occurs without fracturing. Joints are therefore brittle structures.

Question 9.5

This is a sinistral shear zone. The block of rock on which the coin sits has been displaced to the left relative to the other block, as shown by the anticlockwise swing of the foliation in the rock (on both sides of the zone).

Question 10.1

We use Equations 10.1 and 10.2, and compare the post-deformation lengths of the coloured axes in Figure 10.2b with their pre-deformation equivalents on Figure 10.2a.

(a) In Figure 10.2a, the circle has a diameter of 30 mm. The long axis of the ellipse has a length of 45 mm. The extension is given by Equation 10.1:

extension, $e = (l - l_0)/l_0$

$= (45–30)/30$

$= +15/30$

$= +0.5$

From Equation 10.2, the *percentage* extension is given by:

% extension $= e \times 100$

$= +0.5 \times 100\%$

$= +50\%$

The plus sign shows that the line has lengthened during deformation.

(b) The short axis of the ellipse in Figure 10.2b is 20 mm long. Using the same method, % extension = –33%. The minus sign shows that the line has shortened during deformation.

Question 10.2

(a) The angle A′D′C′ in Figure 10.4b is 63°. Therefore, angle ψ, the shear, is 90°–63° = 27°. From Equation 10.3:

$\gamma = \tan \psi$

$= \tan 27°$

$= 0.5.$

(b) As in Question 10.1, use Equations 10.1 and 10.2, and compare the post-deformation lengths of the ellipse axes in Figure 10.4b with the diameter of the circle in Figure 10.4a. The circle in that Figure has a diameter of 30 mm. The long axis of the ellipse in Figure 10.4b has a length of 38 mm. The % extension is given by Equations 10.1 and 10.2:

extension, $e = (l - l_0)/l_0$

$= (38–30)/30$

$= +8/30$

$= +0.27.$

From Equation 10.2, the *percentage* extension is given by:

% extension $= e \times 100$

$= +0.27 \times 100\%$

$= +27\%.$

(c) The short axis of the ellipse in Figure 10.4b is 23 mm long. Using the same method as (b) above, % extension = –23%.

Question 10.3

The long axis of the ellipse is 30 mm and the short axis is 15 mm. The axial ratio of this ellipse is the ratio of long to short axes, i.e. 30 : 15 or 2 : 1.

Question 10.4

The angular shear, ψ, is the *change* in angle between two lines originally at right angles. We can use the bilateral symmetry of the trilobite to give us lines originally at right angles. From Figure 10.5, the angle between the central axis of the fossil and the horizontal body segments, which was originally a right angle, is now about 45°. Therefore, angle ψ, the shear, is 90°–45°=45°, so the shear strain $\gamma = \tan \psi = \tan 45° = 1$. You can't measure the extension of, say the length of the trilobite, because you don't know how long the sheared specimen was *before* deformation (even though you know how long its friend is!).

Question 10.5

You may have chosen to label one of several similar features in Plate 10.1. The hangingwall of the thrust has been translated from right to left, and the individual folded beds show rotation. The fact that once-planar, horizontal beds are now curved and folded is evidence for distortion.

Question 11.1

Say the Alps quadrupled in thickness, from an original thickness t, to a thickness of $4t$. This would be an extension of $(4t - t)/t = 3$, in 10–20 Ma, i.e. a strain rate of between 3 per 10 Ma and 3 per 20 Ma. In (s^{-1}), this is between $3/10(3 \times 10^{13})\ s^{-1}$ and $3/20(3 \times 10^{13})\ s^{-1}$, or between $10^{-14}\ s^{-1}$ and $0.5 \times 10^{-14}\ s^{-1}$. This doesn't even represent one order of magnitude difference from the earlier calculation.

Question 11.2

Normal faults, thrust faults and joints are all brittle structures and are likely to form in the upper parts of the crust. Ductile shear zones are more likely to form at depth.

Question 11.3

If the rock was homogeneous (a), it could simply change its shape uniformly, so that an original cube became shorter in one direction and thicker at right angles to that direction. Rotation of grains and pressure dissolution would produce a slaty cleavage. If the rock was heterogeneous (b), less competent units may shorten this way but more competent units would form buckle folds, with or without a cleavage.

Question 11.4

A parallel fold is one in which the thickness of the folded layer remains constant right around the fold. It implies the layer was relatively competent. Not many layers can form parallel folds before a space problem arises in the fold core. A similar fold is one in which all folded surfaces have the same curvature. They generally form in incompetent, mobile material and require significant volume changes along individual layers. Similar folds can be stacked together indefinitely.

Question 12.1

On the craton, the sedimentary cover would probably be thin. Sediment thickness would increase oceanwards, in the form of a blanket of sediment in the upper part, and irregularly within rotated blocks in the lower part of the basin fill. As oceanic crust was approached, sediments would thin again, tapering to almost nothing on the floor of the ocean proper.

Question 12.2

On the craton, there would be no significant structures. Oceanwards, a dipping but unfaulted blanket of sediment in the upper part would drape normal-faulted blocks in the lower part of the basin fill. These normal faults may dip steeply either oceanwards or landwards. Locally, upfaulted blocks may occur. This structure would be maintained until oceanic crust occurred.

Question 12.3

–30%. The present-day straight-line distance between the ends of the middle bed of the Cretaceous succession, from the WNW end of the section near the Ratz anticline to ESE of the Chartreuse Hills, is approximately 21 km (adjusting for the horizontal scale on the cross-section). Measuring the actual length of that same horizon, around all the folds and across the faults, even where it has now been eroded away, gives a pre-deformation length of about 30 km. That horizon was initially 30 km long – now it is only 21 km long. The percentage extension can now be calculated as per Question 10.1:

extension, $e = (l - l_0)/l_0$ (Eqn 10.1)

$= (21-30)/30$

$= -9/30$

$= -0.3.$

From Equation 10.2, the *percentage* extension is given by:

$\%\ \text{extension} = e \times 100$

$= -0.3 \times 100\%$

$= -30\%.$

ACKNOWLEDGEMENTS

Grateful acknowledgement is made to the following for permission to reproduce material in this Block:

Cover image copyright © Derek Hall; Frank Lane Picture Agency/Corbis; *Figure 4.4a, b* D. Laporte (1994) 'Wetting behaviour of partial melts during crustal anatexis', *Contributions to Mineralogy and Petrology*, **116**, pp. 486–499 © Springer-Verlag GmbH; *Figure 5.1* from *Principles of Geology* (3rd edn) by J. Gilluly, A.C. Waters and A.O. Woodford. Copyright © 1968 by W.H. Freeman, used by permission; *Figure 5.10* G.A. Morris and D.H.W. Hutton (1993) 'Evidence for sinistral shear associated with emplacement of the early Devonian Etive dyke swarm', *Scottish Journal of Geology*, **29**(1), pp. 69–72; *Figure 6.1a–d* T.K.P. Gregg and J.H. Fink (1995) 'Quantification of submarine lava-flow morphology through analog experiments', *Geology*, **23**(1), pp. 73–6, The Geological Society of America; *Figure 6.3* Peter Francis, Open University; *Figure 6.5* Tony Waltham, Geophotos; *Figures 6.6 and 6.21* NASA; *Figure 6.11* Stuart Hall, British Antarctic Survey; *Figure 6.17* US Library of Congress; *Figure 8.11* B.W.D. Yardley (1989) *An Introduction to Metamorphic Petrology*, Addison-Wesley Longman Ltd.; *Figure 8.12* A.E. Maxwell (1970) *The Sea*, by permission John Wiley & Sons Inc.; *Figures 9.1, 9.6, 9.8a, 10.3, 11.1, 11.2, 11.7, 11.10 and 12.7* J.G. Ramsay and M.I. Huber (1983) *The Techniques of Modern Structural Geology*, Academic Press; *Figure 9.5* J.G. Ramsay (1967) 'Shear zone geometry: a review', *Journal of Structural Geology*, **2**(1/2), p.360, with permission of Elsevier Science; *Figures 11.3 and 12.3* Shell Manual; *Figure 11.12* Peter Webb; *Figure 11.14a* J.G. Ramsay; *Figure 12.2* P.A. Ziegler (1982) *Geological Atlas of Western and Central Europe*, Shell International; *Figure 12.4* from an illustration by Ian Worpole in P. Molnar (1986) 'The structure of mountain ranges', *Scientific American*, July, Copyright © 1986 by W.H. Freeman, used by permission.

INDEX

Note: page numbers in **bold** are for terms that appear in the *Glossary* while page numbers in *italics* are for terms contained within Figures.